ALIENS and UFOs: CASE CLOSED
Proof beyond a reasonable doubt

Artist's rendering of a Flying Saucer

Angelo Tropea

Cover by Christopher Tropea

ALIENS AND UFOs: CASE CLOSED / 2

Copyright 2012 by Angelo Tropea.
All rights reserved. No part of this book may be reproduced in any form or by any electronic or mechanical means, including information storage and retrieval systems without permission in writing from the publisher.

Published by Angelo Tropea
P.O. Box 26271, Brooklyn, NY 11202-6271.

ISBN-13: 978-1481107181
ISBN-10: 1481107186

Giordano Bruno

"There are innumerable suns, and an infinite number of earths orbiting around those suns...

We discern only the largest suns, immense bodies. But we do not discern earths because, being smaller, they are invisible to us."

Giordano Bruno, Italian philosopher of the 1500s, burned at the stake for his beliefs

ALIENS AND UFOs: CASE CLOSED

CONTENTS

1. INTRODUCTION / 4
   Evidence In A Court Of Law / 5
   Our Approach In This Book / 7

2. ANCIENT HISTORICAL PERSPECTIVE / 8
   First UFOs and Aliens Sightings / 9
   Historic Aliens / 10
   Julius Obsequens / 11
   Similarity of the Pyramids / 12
   The Cargo Cults / 14
   The Nazca Lines / 15
   Air-chariots Of Ancient India / 16
   Christopher Columbus / 17
   Battle Over Nuremberg / 18
   Religious Testimony Of Our Western Ancestors / 19

3. APPLICATION OF SCIENTIFIC REASONING / 22
   Plurality Of Inhabited Worlds / 23
   Drake Equation / 24
   Kepler Mission / 26
   Habitable Zones On Earth / 27
   "Goldy Locks" Zones / 28
   The Impossible DNA / 30
   The Two Inch Ruler / 32
   Classification Of Sightings And Encounters / 34

4. COMMON SENSE ARGUMENT / 36
   Universe Of The Infant / 37
   New Kids On The Galactic Block / 38

5. RECENT HISTORICAL PERSPECTIVE / 40
   Aurora Incident / 41
   Foo Fighters / 43
   Roswell / 44
   UFOs Over Washington, D.C. / 46
   Betty and Barney Hill / 48
   Japan Air Lines / 50
   Lights Over Phoenix / 51

6. THE TESTIMONY OF EXPERTS / 53
   Astronauts: Scientifically Trained Witnesses / 54
   Carl Sagan and the Ancient Sumerian Language / 56
   The Unknown Universe / 58

7. SELECT LIST OF UFOs IN HISTORY / 59

8. TO YOU, THE JUDGE AND JURY / 91

INTRODUCTION 1

Evidence In A Court Of Law / 5

Our Approach In This Book / 7

Evidence In A Court Of Law

After thirty years of service with the courts, I have become sensitive to the type and quality of proof and the amount of evidence necessary to establish or "prove" a case. Although a case is never irrevocably proved one hundred percent, there are legal standards which allow a case to be proved within prescribed limits of reasonable certainty.

Liberty Holding Scales of Justice

In Family Courts, for example, the evidentiary requirements are less stringent than in Civil or Criminal courts. A Judge may decide a case upon proof that causes the Judge to *reasonably believe* that an event happened.

In Civil Courts, the Judge and Jury apply a higher standard - *preponderance of the evidence* (more than fifty percent probability).

ALIENS AND UFOs: CASE CLOSED

In Criminal Courts, the standard of proof is *beyond a reasonable doubt*. This is the most stringent proof (almost certain that it happened). This of course is necessary because a defendant found guilty is subject to fines and imprisonment.

So what standard of proof should we use to reasonably prove that UFOs and Aliens are just as real as space shuttles and astronauts?

Supreme Court of the United States

To avoid debate, let us apply the most stringent standard – *beyond a reasonable doubt*.

To apply this standard, we need to examine all evidence carefully for three qualities:

1. is the evidence provided by someone who is *competent* (someone who is mature and not mentally impaired?)
2. is the evidence *relevant*? (does it pertain directly to the subject of the inquiry?)
3. is the evidence *material*? (does it have sufficient importance for it to be worthy of our serious consideration?)

And then – does all the admissible evidence as a whole prove the case *beyond a reasonable doubt*?

Our Approach In This Book

In this book we will present reasons (by means of exhibits) why UFOs and Aliens exist and then at the end of the book we will discuss these examples in light of the standards of evidence.

Of course there will always be those who will insist that we should not apply legal standards of proof when discussing UFOs and Aliens. They will insist that the reality of UFOs and Aliens must be proved beyond the shadow of a doubt (with an alien carcass and a wreck of a UFO). They will argue that proving the existence of UFOs and Aliens is "different" from proving a case. If we applied the standards of these doubters to the universe, we would still be thinking of stars as just twinkling lights.

In a court of law murder can be proven with sufficient circumstantial evidence. Witnesses, both regular citizens and also expert witnesses, may testify as to what they saw or what material or other evidence means in their opinion.

For all doubters there must be a point beyond which there is sufficient proof for reality to set in. In terms of the existence of UFOs and Aliens, I think that the reasons in this book will provide the quantity and quality of such proof. All I ask is that you read each section with an open mind, consider the evidence and arguments and then, like a Judge or juror, decide for yourself.

ANCIENT HISTORICAL PERSPECTIVE — 2

First UFOs and Aliens Sightings / 9

Historic Aliens / 10

Julius Obsequens / 11

Similarity of the Pyramids / 12

The Cargo Cults / 14

The Nazca Lines / 15

Air-chariots Of Ancient India / 16

Christopher Columbus / 17

Battle Over Nuremberg / 18

Religious Testimony Of Our Western Ancestors / 19

Exhibit 1
First UFOs and Aliens Sightings

One of the biggest misconceptions of uninformed UFO and Alien skeptics is that sightings of aliens and UFOs are recent phenomena. A careful examination of the historical record makes it clear that UFOs and Aliens have been recorded by our ancestors as far back as the dawn of civilization – by the Sumerians, Hittites and Egyptians. Prior to that there are examples in cave drawings that something other than worldly creatures are being depicted.

Ancient Cave Drawings

These observations continued uninterrupted to the present time when each sighting may be communicated to the world in a matter of minutes. I wonder how many more sightings would have been recorded by our ancestors if they knew how to write or had the needed understanding to realize what they were seeing. Much knowledge and proof may thus have evaporated by the passage of time.

But all this changed as people became literate and able to record their experiences and thereby communicate with their contemporaries and their descendants. In the following pages we will attempt to relate some of the written testimonial records of our ancestors.

Exhibit 2
Historic Aliens

In recent years UFO and Alien researchers have been working hard to prove the theory that the earth has been visited in both historical times and in prehistory by beings from other worlds. The evidence has been steadily accumulating and has now reached a point where it is difficult to summarily dismiss the theory.

One of the most famous proponents of this theory is author and researcher Erich von Daniken. For more than five decades he has been writing about the connection among ancient architecture, artifacts, art and ancient alien visitors.

Beyond the depiction in ancient art of beings that resemble astronauts and alien creatures and space vehicles, he points to massive buildings such as the pyramids and mysterious monumental artwork such as the Nazca lines in Peru as projects that were well beyond the expected artistic and architectural limits of their societies.

In addition to the artistic and architectural connections, some UFO and Alien researchers also point to the common thread among many religions of visitors from the sky descending inside fiery objects. These threads are shared in oral and written accounts by religions in both the European, American, Asian, African and Australian traditions.

Of course not all researchers and scientists share this ancient aliens theory. Carl Sagan, a respected scientist, was of the opinion that many of the "examples" of alien contact in ancient times could upon close inspection be easily explained without the need for alien intervention. Carl Sagan died before the current discoveries of many worlds. I wonder if he might have modified his thinking if he were living today.

Exhibit 3
Julius Obsequens

Julius **Obsequens was a Roman author** of the fourth century A.D. He is best known for his work *Liber de prodigiis,* based on a written work by the historian Livy (59 BC – AD 17). The book describes unusual occurrences that happened in Rome between the years 249 BC – 12 BC.

The following are three passages cited in the Wikipedea article "Julius Obsequens."
1. (100 BC): "When C. Murius and L. Valerius were consuls, in Tarquinia towards sunset, a round object, like a globe, a round or circular shield, took its path in the sky from west to east."
2. (91 BC): "At Aenariae, while Livius Troso was promulgating the laws at the beginning of the Italian war, at sunrise, there came a terrific noise in the sky, and a globe of fire appeared burning in the north. In the territory of Spoletum, a globe of fire, of golden color, fell to the earth gyrating. It then seemed to increase in size, rose from the earth and ascended into the sky, where it obscured the sun with its brilliance. It revolved toward the eastern quadrant of the sky."
3. (42 BC): "something like a sort of weapon, or missile, rose with a great noise from the earth and soared into the sky."

While some skeptics might dismiss the first observation ("took its path from east to west") as possibly a meteor, they cannot use the same reasoning for sightings 2 and 3. Meteors do not rise from the earth and ascend into the sky.

We should also note that Obsequens does his best to use his limited contemporary knowledge to describe the flying objects: "round object...like a globe...round or circular shield, globe of fire...gyrating... like a sort of weapon or missile (projectile)."

Exhibit 4
Similarity Of The Pyramids

Almost everyone would agree that the greatest **architectural monuments** built by ancient humans are the gigantic pyramids. Huge in scope and awesome to look at, each was built by thousands of people toiling for years and perhaps decades. Almost all were constructed for religious and burial (afterlife) reasons and all shared a common shape – that of the pyramid.

The similarity of shape of these monuments in Egypt, Asia, and in the southern American Hemisphere is mind boggling when one considers the vast distances among these areas and the impossibility of these civilizations communicating with each other. In addition to speaking different languages, there is no record of any of these people travelling to the other areas at the times these structures were built. The Egyptians never strayed further than Africa and the Middle East. The Aztecs, Mayans and Incas of South America travelled on foot and were not even aware of the wheel. And incredibly distant Asia at the time that the mound pyramids were built was impossible to reach. Furthermore, the base of the pyramid at Giza surprisingly matches the size of the one at Teotihuacan (City of the Gods) in present day Mexico.

In addition to a common shape, there is a startling similarity of configuration of the pyramid complexes at Teotihuacan and those at Giza (Egypt). Both configurations, or layouts of the three main pyramids, match the configuartion of the three middle stars of the constellation of Orion. This has been termed the "Orion correlation theory." This theory postulates that the correlation was intended by the builders of the pyramids and that this claim is bolstered by the fact that the Orion stars were linked to the Egyptian god Osiris (the god of rebirth and afterlife – concepts which also heavily influenced the religion of the builders of the pyramids at Teotihuacan).

The Great Pyramid at Giza, Egypt

The Pyramid of the Sun at Teotihuacan, Mexico

Exhibit 5
The "Cargo Cults"

To provide an example of how religions based on the **visitation** of ancient aliens could have developed, some researchers point to the "Cargo Cults" of the South Pacific during the twentieth century. Prior to these cults there existed the Tuka Movement and the Taro Cult during the 1800s.

The common element among cargo cults was the visitation of people (Westerners) with superior scientific and technical knowledge who "wowed" the natives with their technical superiority and physical possessions, including clothing made of fabric that seemed constructed by magical means, weapons of great power, and food that was strange and new and very much desired.

During the second world war, Japanese soldiers and then Allied forces occupied a number of the Melanesian islands in the Pacific. Both forces had soldiers stationed on the islands. They frequently received weapons, food and equipment. These supplies were dropped to the soldiers from cargo planes. The primitive natives witnessed not only the fighting, but also the provisioning of the soldiers who often shared their food with their cooperative and appreciative hosts.

After the war ended, the soldiers left and the flood of food and supply of seemingly magical equipment ceased. Soon some of their religious leaders promoted cults based upon the visitation of the visitors from the sky. Their members developed religious rituals, aimed at prompting the return of the visitors and their material wonders. Some of their rituals mimicked the behavior of the soldiers, such as marching with wooden rifles and establishing runways with mock airplanes constructed of wood and straw. As the years passed and the visitors didn't return, the cults waned. Today with modern communication and knowledge, many of the islanders realize that the visitors were not gods, but much more technically advanced people than them.

Exhibit 6
The Nazca Lines

An evidentiary example of an ancient "Cargo Cult" may be the Nazca Lines of Peru. Some of these gigantic line drawings on the ground of a high and arid plateau in the Nazca Desert date as far back as 200 B.C. (with some as recent as 700 A.D.). The people that drew these line drawings of animals and geometric figures did so for almost one thousand years – and not for their own enjoyment (The drawings were so large, some up to 600 feet, that the shapes can only be discernible from high above in the sky).

Monkey line drawing

Some of these lines are straight and wide, reminiscent of airplane runways. Why would the Nazca people make these lines? Were they trying to send a message to someone who could see them from above? Lacking a written language and the ability to transmit messages by other means, isn't it reasonable to ask the question – Were the Nazca people asking someone to visit them – or possibly return?

Exhibit 7
Air-Chariots of Ancient India

In ancient Indian epics such as the *Ramayana* (5th or 4th century B.C.) and in the more ancient *Mahabharata*, the gods travel in "flying cars" or "flying chariots," called *Vimanas*.

> From the *Ramayana* Book 6, Canto CXXIII: The Magic Car:
> *This chariot, kept with utmost care,*
> *Will waft thee through the fields of air,*
> *And thou shalt light unwearied down*

> From Book 6, Canto CXXIV: The Departure:
> *Swift through the air, as Rama chose,*
> *The wondrous car from earth arose.*
> *And decked with swans and silver wings*
> *Bore through the clouds its freight of kings.*

> From the *Mahabharata*:
> *Bhima flew with his Vimana on an enormous ray which was as brilliant as the sun and made a noise like the thunder of a storm.*

If we put ourelves in the shoes of the ancient writers of the *Ramayana* and the *Mahaharata*, we can understand their use of contemporary terms such as "chariot" and "car" to describe a flying vehicle, "decked with swans and silver wings" to describe engines, and "enormous ray" and "noise like thunder" to describe rockets igniting and vehicle exhaust.

Exhibit 8
Christopher Columbus (1492)

Almost everyone knows about Christopher Columbus and the New World, but few know that on October 11, 1492, just hours before land was sighted, Christopher Columbus saw a UFO, an incident which he recorded in his journal.

While on the deck of the Santa Maria, Columbus spotted "a light glimmering at a great distance." It moved up and down. When Columbus saw the unexplained light, he called for one of the sailors, who also saw the light which vanished after a while, only to reappear several more times during that fateful night.

Christopher Columbus

Upon the return to Spain, the sailor was very concerned about the sighting and remarks by Christopher Columbus which he considered heretical. He reported Columbus to the Spanish Inquisition who questioned Columbus about what he saw and his loyalty to the Christian faith. Luckily Columbus was able to come out of the investigation unscathed. The accusation and subsequent investigation by the Inquisition add support to the reality of the incident and the backlash that it had at the time.

It is important to note that Columbus didn't have to include the sighting in his journal. He could just as well have left it out and probably no one would have noticed. But he did include it in his journal, obviously thinking that it was essential for an accurate historical record.

Exhibit 9
Battle Over Nuremberg (1561)

If you had not read the **title of this section** and just examined the drawing below (from a woodcut allegedly depicting the details from a news notice reporting the strange celestial events of April 4, 1561, in Nuremberg, Germany), what would you say it attempts to depict?

It is clear that the bottom portion depicts a landscape,

Print From A Woodcut

with a town on the left, including a church with two steeples, and a number of plowed fields adjacent to the town. But what are the objects above it - in the sky?

The large circle near the center is clearly the shining sun, depicted in the traditional way - with eyes, a nose, and a mouth. But what are the other various sized circles and cylindrical objects all around it?

According to one translation, numerous men and women saw two opposing factions battling against each other for two hours. "The Gazette" reported *"...the dreadful apparition (which) filled the morning sky with cylindrical shapes from which emerged black, red, orange and blue-white spheres that darted about. Between the spheres there were crosses with the color of blood."*

It is interesting to note that several of the "different colored" spheres had apparently crashed down (bottom right of the woodcut) and that they were in flames (the two smoke clouds rising from the crashed spheres).

The large, black spear-like shape is unlike the others, both in size and color. Some theorists interpret it as a mother ship.

Exhibit 10
The Religious Testimony
Of Our Western Ancestors

Ancient religious texts from the West also have passages which describe other worldly beings and magical vehicles.

Examples of such passages may be found in the *Book of Genesis,* the apocryphal *Book of Enoch,* and the *Book of Ezekiel.*

> **From the Book of Genesis, chapter 6, verses 1-4:**
> *When human beings began to increase in number on the earth and daughters were born to them, <u>the sons of God</u> saw that the daughters of humans were beautiful, and they married any of them they chose.*

ALIENS AND UFOs: CASE CLOSED / 20

From the Book of Ezekiel, chapter 1:

"And I looked, and behold, a whirlwind came out of the north, a great cloud, and a fire unfolding itself, and a brightness was about it, and out of the midst thereof as the color of amber, out of the midst of the fire.

(verse 5) Also out of the midst thereof came the likeness of four living creatures. And this was their appearance; they had the likeness of a man.

(verse 19) And when the living creatures went, the wheels went by them: and when the living creatures were lifted up from the earth, the wheels were lifted up (verse 20) Whithersoever the spirit was to go, they went, thither was their spirit to go; and the wheels were lifted up over against them: for the spirit of the living creature was in the wheels.

(verse 21) When those went, these went; and when those stood, these stood; and when those were lifted up from the earth, the wheels were lifted up over against them: for the spirit of the living creature was in the wheels.

While reading and contemplating the meaning of the passages from the *Ramayana* and *The Book of Ezekiel,* it is important to keep in mind that the writers of the time didn't know about aliens, space vehicles, rockets or fuel exhaust. These words didn't exist for them. It is fair to think that they drew upon their limited experience and vocabulary to describe unknown beings as *sons of God,* and vehicle exhaust as *"a whirlwind (that) came out of the north, a great cloud, and a fire infolding itself, and a brightness..."*

How would Ezekiel describe beings travelling in a space vehicle? The following passage would seem a good attempt, based on the technology of his time (593 B.C.)

(verse 19) And when the living creatures went, the wheels went by them: and when the living creatures were lifted up from the earth, the wheels were lifted up.

Ezekiel then continues to emphasize the transportation of the creatures by the "wheels" in verses 20 and 21 (see preceding page).

The above examples draw from written testimonies. Prior to the development of writing, depictions of aliens and their visitation may be found in Paleolithic cave paintings (see illustration on page 6). Both Medieval and Renaissance art also provide examples of UFOs.

In summary, UFO and alien reports date as far back as there is a written history, either in art or in written accounts – way before the first helicopter or airplane ever flew, and way before the first helium balloon rose into the sky.

APPLICATION OF SCIENTIFIC REASONING | 3

Plurality Of Inhabited Worlds / 23

Drake Equation / 24

Kepler Mission / 26

Habitable Zones On Earth / 27

"Goldy Locks" Zones / 28

The Impossible DNA / 30

The Two Inch Ruler / 32

Classification Of Sightings And Encounters / 34

Exhibit 11
Plurality Of Inhabited Worlds

Giordano Bruno

Giordano Bruno (1548 – 1600) was an Italian philosopher, mathematician, astronomer and Dominican friar. In the year 1600 he was found guilty of heresy and burned at the stake. His crimes included a belief that the sun was itself a star and that the universe was populated with countless stars with an infinite number of worlds inhabited by intelligent beings (a belief referred to as the "plurality of inhabited worlds").

In this and other beliefs, Giordano Bruno was ahead of his time; the telescope had only recently been invented and the majority of the people of the world were still illiterate. In addition, he lived in the time of the inquisition, when scientific thought in opposition of Church teachings was severely persecuted.

Although given the opportunity to recant and thereby save his life, Giordano Bruno refused to do so. His belief that many worlds existed did not die with him. But it was only in our twenty-first century that scientists were able to confirm the existence of many worlds beyond our solar system.

Giordano Bruno

Exhibit 12
Drake Equation

From Bruno's sixteenth century to present day other great thinkers continued Bruno's quest for knowledge. One such person is Frank Donald Drake, an American astronomer and astrophysicist and founder of SETI (Search for Extraterrestrial Intelligence). An equation which he developed in 1961 estimates the approximate number of intelligent civilizations in our galaxy, the Milky Way.

$$N = N * Fp\ Ne\ Fi\ Fi\ Fc\ FL$$

The following lists an example of "best guess" values for each variable in the equation. As in all estimates, the values and the result may vary, depending on who is estimating.

> N = the number of stars in our galaxy (the Milky Way)... 100 billion
>
> Fp = the fraction of stars that have planets around them... up to 50% (recently updated to almost 100% !!!)
>
> Ne = the number of planets around each star that are able to support life... up to 5
>
> F1= the fraction of planets in Ne where life evolved...any percentage up to 100%
>
> Fi = the fraction of Fl where intelligent life evolved...any percentage up to up to 100%
>
> Fc = the fraction of Fi that is able to communicate...any percentage up to 20 %
>
> FL = the fraction of the existence of the planet during which civilization that can communicate live... as low as one millionth of the life span of the planet (in the Earth's case, an estimate of 10,000 years)

Based on this equation and applying very conservative values, the number of communicating civilizations in our galaxy alone may number in the thousands. The Drake Equation proved to be a valuable tool for more than fifty years.

> An equation is a set of numbers and letters, often unassuming at first glance. Its significance and power can only be appreciated with careful examination and consideration.
>
> Consider the power of the following seemingly simplistic and unassuming equation devised by a clerk (Albert Einstein) at a European copyright office:
>
> $$E = MC^2$$

Exhibit 13
Kepler Mission

Recently another tool has been added in the search for extra terrestrial worlds: the Kepler Space Mission. Its aim is to find alien worlds, including other Earth-like planets, in a distant patch of the Milky Way galaxy.

Recent findings from the Kepler mission have added to other findings from earth based scientists that have been searching for alien worlds through a variety of innovative methods, including the *transit method* (which measures the shadows of planets passing in front of their stars) and the *radial velocity* and *astrometry* method (which measure the wobble of a star caused by the planet's gravitational tug).

NASA's Kepler Telescope

These methods have already helped to discover hundreds of planets and have encouraged scientists to estimate that there may be more than 100 billion planets in our galaxy alone.

But how many of the planets are earth-like? And how many of the earth-like planets have inhabitants who can communcate with us? The number of possibilities is growing every day as new planets are found and the environmental boundaries within which like can exist expand. As earth based studies and outer space scientific observation gives us the basis to increase the "guess" numbers in the Drake equation, the number of probable planets – including earth - like planets - increases dramatically.

Exhibit 14
Habitable Zones On Earth

Until recently, before the discovery of deep sea thermal vents and before enlightened research into boiling geysers, mines and other environments considered impossible for life, the hospitable range for life (temperature and other conditions, such as the presence of water and sunlight) was very narrow. Because of this the possibility of life on other worlds was considered small.

But scientists have recently been surprised to discover microbial organisms on earth (called extremophiles) thriving around boiling deep sea vents, in acidic hot puddles, in dry environments such as salt flats, and in mines with little heat and absolutely no sunlight. The range of conditions within which life can exist has been widening to

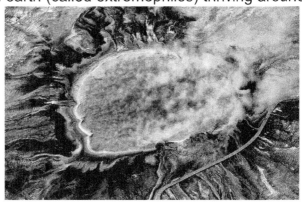

Thermophiles, a type of extremophile

the point where some scientists hypothesize that the conditions which we once thought were essential for life (including the presence of water and sunlight) may not be essential to every type of life.

It is interesting to note that mankind has been sharing the earth for millions of years with extremophiles without having the smallest clue of their existence. This is an example of our severely primitive understanding of nature and the world around us. Considering this, should we be surprised that we may perhaps have an equivalent primitive understanding of other life in the universe?

Exhibit 15
"Goldy Locks" Zone

E**ven assuming that water,** moderate temperature and sunlight are essential for life to exist and evolve, there are countless areas around stars where these conditions exist. Scientists calculate that around each star there is an area that is a proper distance from that star where a planet with atmospheric pressure can maintain liquid water and a habitable temperature. Scientists call these zones "habitable zones."

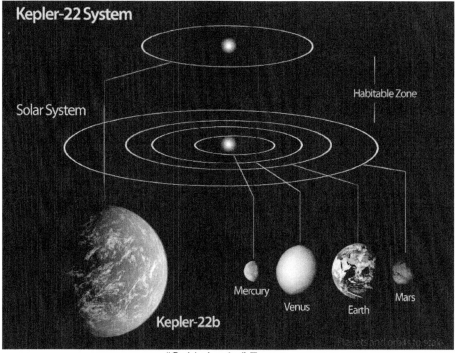

"Goldy Locks" Zones

Some refer to these zones as "Goldilocks Zones" – where the temperature is not too cold or too hot and all the other conditions for life as we know it are "just right."

According to NASA, "This diagram compares our solar system to Kepler-22. Kepler-22's star is a bit smaller than our sun, so its habitable zone is slightly closer in. The diagram shows an artist's rendering of the planet comfortably orbiting within the habitable zone, similar to where Earth circles the sun."

But does life – either microbial or advanced – have to be restricted to our comfort zones? Why can't there be life that breathes methane instead of oxygen, or that thrives in cold or heat instead of our thin range of tolerable temperatures? If we don't limit "habitable zones" to those that only support life as we know it, then the possibility of life in the universe turns from a possibility to a probability, especially considering that the surprising number of recently discovered planets in just a tiny selected speck of our galaxy.

If scientists are correct that there are more than one hundred billion planets in our galaxy, how many planets are there in the more than 100 billion other galaxies in our universe? The numbers are staggering and the possibility of life is so great that we may think of it as the certainty of life.

Relative size of first five planets discovered by Kepler Telescope

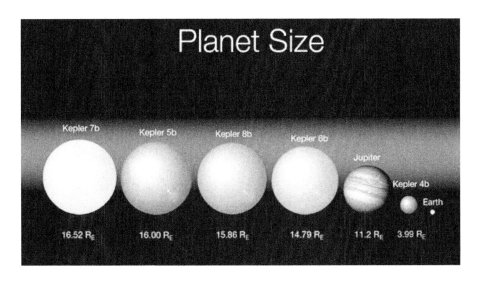

Exhibit 16
The Impossible DNA

Professor Francis Crick (1916 - 2004), the co-discoverer of the double helix structure of the DNA molecule, proposed the *theory of directed panspermia* in 1973. For Professor Crick, this theory offers an interesting explanation for the origin of life and mankind.

DNA Double Helix Structure

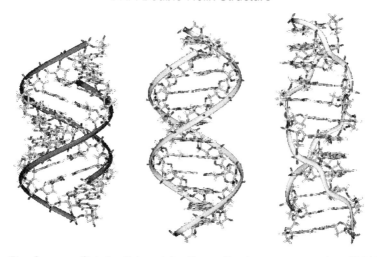

Professor Crick did not believe that very complex DNA could evolve by itself on earth in the short geologic span of time that it did. He stated, "You would be more likely to assemble a fully functioning and flying jumbo jet by passing a hurricane through a junk yard than you would be to assemble the DNA molecule by chance. In any kind of primeval soup in 5 or 600 million years, it's just not possible."

The Panspermia theory hypothesizes that life took root in another place and time in the universe, way before life on earth took hold, and that life on earth and throughout the remainder of the universe was distributed by meteoroids, planetoids and asteroids. A discovery which supports this theory is the discovery that earth bacteria have been able to survive on the moon and on NASA satellites. This was a discovery made due to the accidental contamination of NASA spacecraft before launch and the hardiness of the bacteria that participated in an unexpected ride.

Another example of a life form "piggy-backing" its way to other worlds is the Mars meteorite ALH84001. Although scientists are still arguing about the results, many scientists believe that the "worm-like" form in the image below is indeed a fossilized life form.

Mars meteorite ALH84001

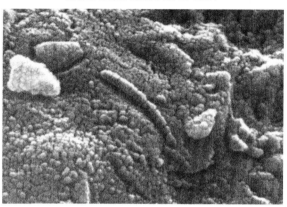

The distributed life forms (such as bacteria, spores and extremophiles) crash on different worlds and entrench themselves. Professor Crick believed that this would explain the sudden outburst of life on earth and the absence of transitional evolutionary forms in the geologic record.

The Theory of Directed Panspermia, proposed by Professor Crick and Leslie Orgel, goes one step beyond the basic *Panspermia theory* by hypothesizing that the seeds of life were intentionally spread by advanced extraterrestrial civilizations. Like the *Panspermia Theory*, it does not preclude the possibility that life may have originated more than once in the universe.

Both the *Panspermia Theory* and the *Theory Of Directed Panspermia* support the existence of UFOs and Aliens, for both argue for the existence of life throughout the universe. Also, both hypothesize that life began much earlier than life on earth, giving that life the time to advance much more than us. Also, both theories provide the possibility that some aliens with similar DNA from us may indeed have one head, two arms and two legs.

Exhibit 17
The Two Inch Ruler

A common saying regarding the limitations of knowledge is "You can't measure the ocean with a two inch ruler." Anyone measuring the ocean with such a ruler would be smug with the knowledge that the depth had been measured and that the science of oceanography would rightly be the science of two inches of salt water.

If we look carefully at the historical record we see many examples of when human beings tried to "measure" or understand the world around them with their limited and often narrow-minded (sociological or religious) understanding.

For centuries our ancestors, including the Greeks of classical times and later the Romans and other societies that followed them (including European, Asian, African and Indians of the Americas) concluded that the earth was at the center of all things and that the sun, like the planets and stars, revolved around the earth (which by careful observation was proven to be flat).

Some of our ancestors hypothesized that everything was made up of four material elements (Earth, Water, Air and Fire). Some added a fifth element, a non-material element called *Aether* (in ancient India and Greece).

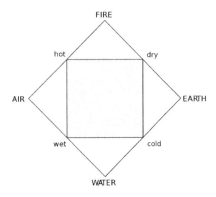

The Four Elements

ALIENS AND UFOs: CASE CLOSED / 33

The four (or five element) theories attempted to describe matter with words and concepts that were familiar and understandable to the people of their times. Although simple and incorrect (an example of the application of the two inch ruler), this limited five elemental theory was an improvement from other primitive attempts to describe matter as magical or caused by actions of the gods.

I wonder if the ancient Greeks or Indians considered their elemental theories to be anything other than perfect. I wonder whether any of the people or scientists or philosophers of those times ever considered the possibility of the more than 100 elements in the periodic table – or the existence of molecules, atoms, protons, electrons and the other sub atomic particles already identified and the countless other particles yet to be discovered?

The Periodic Table Of Elements

Group → Period ↓	1	2	3	4	5	6	7	8	9	10	11	12	13	14	15	16	17	18
1	1 H																	2 He
2	3 Li	4 Be											5 B	6 C	7 N	8 O	9 F	10 Ne
3	11 Na	12 Mg											13 Al	14 Si	15 P	16 S	17 Cl	18 Ar
4	19 K	20 Ca	21 Sc	22 Ti	23 V	24 Cr	25 Mn	26 Fe	27 Co	28 Ni	29 Cu	30 Zn	31 Ga	32 Ge	33 As	34 Se	35 Br	36 Kr
5	37 Rb	38 Sr	39 Y	40 Zr	41 Nb	42 Mo	43 Tc	44 Ru	45 Rh	46 Pd	47 Ag	48 Cd	49 In	50 Sn	51 Sb	52 Te	53 I	54 Xe
6	55 Cs	56 Ba		72 Hf	73 Ta	74 W	75 Re	76 Os	77 Ir	78 Pt	79 Au	80 Hg	81 Tl	82 Pb	83 Bi	84 Po	85 At	86 Rn
7	87 Fr	88 Ra		104 Rf	105 Db	106 Sg	107 Bh	108 Hs	109 Mt	110 Ds	111 Rg	112 Cn	113 Uut	114 Uuq	115 Uup	116 Uuh	117 Uus	118 Uuo

Lanthanides	57 La	58 Ce	59 Pr	60 Nd	61 Pm	62 Sm	63 Eu	64 Gd	65 Tb	66 Dy	67 Ho	68 Er	69 Tm	70 Yb	71 Lu
Actinides	89 Ac	90 Th	91 Pa	92 U	93 Np	94 Pu	95 Am	96 Cm	97 Bk	98 Cf	99 Es	100 Fm	101 Md	102 No	103 Lr

I wonder if the people who dismiss the existence of UFOs and Aliens are doing so by applying their personal "two inch ruler."

Exhibit 18
Classification Of Sightings and Encounters

In an attempt to better describe UFOs and aliens sightings and encounters, a number of researchers developed classification scales. One such researcher was astronomer J. Allen Hynek. His threefold classification scale was as follows:

1. Encounters of the FIRST kind (Sightings of a UFO)

2. Encounters of the SECOND kind (Sightings of a UFO, with the encounter causing a physical effect on animate or inanimate objects)

3. Encounters of the THIRD kind (Sighting of UFO plus its occupants)

Researchers extended Hynek's scale, with some categorizing encounters up to the SEVENTH kind.

Although these categories help classify the encounters, some argue that the categories plus the publicity created by the media, including with such films as "Close Encounters of the Third Kind," fuels the imagination of the public and increases the number of reported sightings. These critics, I believe, fail to seriously consider the sightings of UFOs and aliens that occurred prior to any categorization of them, with many sightings occurring before J. Allen Hynek or any of his critics were born.

But why then are we describing the classification of sightings as evidence of the existence of Aliens and UFOs? Not because of the classification, but because of the person who pioneered the classification – J. Allen Hynek.

J.Allen Hynek (1910 – 1986) an American astrophysicist, astronomer and professor (Ohio State and Northwestern University) served as a U.S. Air Force scientific advisor in three UFO studies:
    1. Project Sign (1947-1949)
    2. Project Grudge (1949-1952)

3. Project Blue Book (1952-1969)

Initially Professor Hynek was skeptical about UFOs. In 1948 he is reported to have said, "the whole subject seems utterly ridiculous." However, as the years passed and he examined numerous UFO and Alien reports, many of which could not be explained, his attitude changed. During the later years of Project Blue Book (1952-1969) he voiced his disagreement with Air Force conclusions regarding some encounters, including the Portage County UFO chase and the encounter of the third kind by police officer Lonnie Zamora near Socorro, New Mexico.

In 1977, at an International UFO Congress in Chicago, Professor Hynek stated, "I do believe that the UFO phenomenon as a whole is real...I hold it entirely possible that technology exists, which encompasses both the physical and the psychic, the material and the mental. There are stars that are millions of years older than the sun. There may be a civilization that is millions of years more advanced than man's. We have gone from Kitty Hawk to the moon in some seventy years, but it's possible that a million-year-old civilization may know something that we don't...."

...a startling statement by the same man who at first thought that "the whole subject seems utterly ridiculous."

Professor Hynek and another person have something in common. That person is Nick Pope. From the age of 20 in 1985 and for the next 21 years, Nick Pope worked for the British Government's Ministry of Defense. From 1991 to 1994, he investigated UFO sightings for the Ministry, and like Professor Hynek became increasingly skeptical that conventional explanations of UFO and Alien reports sufficiently explained all of the reports.

It is interesting to note that two high level investigators in responsible governmental positions reached similar conclusions.

COMMON SENSE ARGUMENT   4

Universe Of The Infant / 37

New Kids On The Galactic Block / 38

Exhibit 19
Universe Of The Infant

An infant lies in his crib. The ceiling light captures his attention and he looks up, trying to understand it.

Between the crib and the light in the ceiling, strange shapes turn – a battery operated carousel of plastic grey elephants, yellow giraffes, blue birds and brown teddy bears. The infant doesn't understand the shapes, but they interest him nonetheless. His eyes follow their movements, all the while anticipating their circular route. The baby doesn't know yet about batteries – or even the nature of colors. He probably isn't even wondering who the provider of the universe above his head might be.

We haven't found any record of what our cave dweller ancestors might have been thinking as they looked up at the sky. For millions of years they didn't record any ponderings or conclusions. It is only during the last few thousand years that they started to scratch the walls of caves and add red ochre (a naturally occurring red pigment) to highlight the shapes.

The ability to communicate by writing developed slowly. As it did, incidents of UFO sightings and alien contact filtered into the written record. To understand this record we have to keep in mind that our ancestors didn't know about science and thought that the world was influenced by demons and magic. We have to wade through ancient mythology and historical accounts with the eye of an interpreter, reading a word and substituting another from a language that we understand and can appreciate.

With this in mind, we will consider selected historical examples of UFOs and Alien contact.

Exhibit 20
New Kids On The Galactic Block

Scientists often like to use the example of a 24 hour clock to explain the relatively short span of time that homo sapiens have existed on earth. At its start (midnight) the earth could be described as a boiling furnace with an atmosphere of poisonous fumes and with meteorites frequently bombarding the surface. Because of these inhospitable conditions, scientists think that it is reasonable to conclude that life (microbial and higher forms) was only possible after the first billion years (during the past 3 billion years).

If we use the 24 hour clock as an analogy to the last four billion years, then from midnight, when the earth formed, to 3:00 a.m. the earth was bombarded with streams of meteorites. The earliest fossils found by archeologists are from 5:36 a.m., but it only around 2:08 p.m. (in the afternoon) that single celled algae are discernible in the fossil record. Seaweed, jellyfish and Trilobites thrived from about 8:30 p.m. to about 9:00 p.m. Plants on land appeared at 9:52 p.m. Dinosaurs walked the earth at 10:56 p.m. and it was not until 11:39 p.m. that mammals made their debut.

So, when did humans first appear? Only at 11:58 p.m. - 2 minutes before midnight! In other words, out of 1,440 minutes in the 24 hours, mankind has been around for for only the last 2 minutes!

During these last two minutes homo sapiens struggled to lift themselves from scattered unclothed hunter gatherers to the most prominent species on earth, who control not only their destiny but also many of the forces of nature. During these last two minutes we have either visited or occupied almost every ecological niche – on the surface, underground and in the sky. We fly like birds, swim like dolphins and we are able to live underground, either in naturally formed caves or in tunnels that we have dug out.

And yet for all our greatness we have to keep in mind that humankind has only been around for two minutes. We have learned much in those two minutes. But will our accumulated knowledge pale in comparison to what we still have to learn in the remaining countless minutes?

We also should keep in mind that simply because we at this time aren't able to travel to the planets or other stars, that doesn't mean that other civilizations – perhaps much older than ours – have not done so. Isn't it highly arrogant and delusionary to think that in the span of two geologic minutes humans have reached the pinnacle of development?

Artist's Rendering Of An Alien City

Isn't it possible that other civilizations reached our stage of development decades, centuries, millennia – or perhaps even a billion years ago? With our elementary knowledge, who are we to think that aliens are not able to warp space, travel faster than the speed of light or journey to other dimensions?

RECENT HISTORICAL PERSPECTIVE   5

Aurora Incident / 41

Foo Fighters / 43

Roswell / 44

UFOs Over Washington, D.C. / 46

Betty and Barney Hill / 48

Japan Air Lines / 50

Lights Over Phoenix / 51

Exhibit 21
The Aurora Incident

Six or seven years before the first flight by the Wright Brothers, many cigar-shaped UFOs were sighted across the United States.

On April 19, 1897, the Dallas Morning News reported an incident at Aurora, Texas, where a UFO was alleged to have crashed with a windmill (now determined to have been a sump pump) owned by Judge J.S.Proctor. According to the report, written by S.E.Haydon, an Aurora resident, a pilot "not of this world" was found dead and mangled and was buried by the townspeople at the Aurora Cemetery. The wreckage of the vehicle, which resembled a mixture of aluminum and silver, and which contained strange "hieroglyphic" images, was allegedly dumped in a well located at the crash site.

Plaque At Aurora Cemetery

Mr. Brawley Oates purchased the property in 1945 and cleaned out the well and used it as a water source. Strangely, Mr. Oates developed a severe and horribly disfiguring case of arthritis which he attributed to the contaminated well water. Because of this, Mr. Oates sealed the well with a concrete slab.

The Aurora incident was investigated by Dallas TV station KDFW, UFO Files, and UFO Hunters. All investigations were inconclusive. The well water tested high for aluminum and the cemetery plot seemed to have been disturbed, now providing no evidence of an alien internment. KDFW reported that Texas had erected a historical plaque which includes the Aurora legend.

The Aurora Incident by itself is interesting but not conclusive. However, when considered with other UFO and Alien reports prior to this incident (in the time *before* airplanes), the Aurora report takes on more weight. Consider the following from Wikipedia:

November 18, 1896: The *Sacrament Bee* and *San Francisco Call* report residents seeing a light in the sky with a shape behind the light moving slowly.

November 19, 1896: The *Daily Mail* reported the account of a Colonel H.G. Shaw who claimed to have seen a landed aircraft (25 feet diameter and 150 feet long) and three slender, 7 feet tall aliens who attempted to pull him into the aircraft. They failed because they lacked the necessary strength and thereafter sped back to their ship and took off.

November 21, 1896: The cities of Sacramento, Folsom, San Francisco and Oakland reported mystery lights in the sky, seen by hundreds of people.

The "Aurora" name coincidence

One of the most widespread rumors regarding United States secret aerial projects has been labeled the "Aurora Project," supposedly being carried out in top secret Area 51. By coincidence, the name of the project is the same as Aurora (Texas) where in 1897 a metal object crashed from the sky.

Exhibit 22
Foo Fighters

Years before the word UFO was introduced into the English language, Allied aircraft pilots during World War II encountered strange and unknown aircraft which they called "foo fighters." Although pilots and their crews reported these sightings formally and their reports were treated seriously, almost nothing was learned about these unidentified aerial phenomena during the war.

"Foo fighter" reports were filed in both the European and Pacific Theaters. Some reports described these crafts as "balls of fire" which sometimes just hung in the sky while at other times they followed the aircraft. One aircraft managed to hit a "foo fighter" with gunfire, causing the foo fighter to explode, its pieces raining down to the ground where they set buildings on fire.

Theories as to what the "foo fighters" might have been range from natural phenomena such as *ball lightning* to German secret weapons – or perhaps to our current understanding of an extra-terrestrial UFO. It is probable that if the term "UFO" had been known at the time, the pilots would have labeled them as UFOs.

These aircraft, whatever they were, at times followed Allied airplanes for more than an hour and caused the steel nerves of airplane pilots to melt. To summarily attribute these widespread reports to pilot error or war fatigue is too convenient and closed minded.

Ball Lightning

Exhibit 23
Roswell

In the summer of 1947 an object, allegedly an aircraft, crashed on a ranch near the town of Roswell, New Mexico. What the object was and where it came from has been the subject of heated public controversy ever since.

Roswell "Daily Record" Front Page

Pressured by UFO investigators, the United States Armed Forces announced that the object was nothing but debris from a weather balloon. Since this did not square with its original announcement, not everyone found this assertion convincing. The public release by an army information officer dated July 8, 1947 stated that the army had recovered a "flying disk." This was immediately reported by the media and generated a great amount of public interest. However, the following day the Air Force, represented by Commanding General Roger M. Ramey, backtracked and identified the fallen aircraft as a radar-tracking balloon. This succeeded in keeping the incident from the public light until 1978, when Stanton T. Friedman, a physicist and UFO investigator, made public his interview with Major Jesse Marcel who had handled the debris of the Roswell UFO.

In the following years, attempts by the United States Armed Forces to explain away the crash only succeeded in intensifying public awareness and interest and in prompting other witnesses and participants of the 1947 crash to come forward. These witnesses included Glenn Dennis (mortician) who stated that alien autopsies were done at the military base at Roswell. The military tried explaining away the sighting of "bodies" at the crash site by stating that anthropomorphic dummies had been used in military tests. However, the UFO community dismissed these claims because according to them such dummies were not used by the military until years later. Some UFO investigators also believe that the debris (which had not been allowed to be closely inspected by the media) and alien bodies were whisked away to air basis for further study.

Some assert that this incident is a prime example of the mishandling of UFO incidents by the military. They claim that through action or inaction, the military has succeeded in creating a cloud of doubt over reports that would otherwise have firmly established the existence of aliens and UFOs.

Exhibit 24
UFOs Over Washington, D.C.

In 1952 there were a startling number of UFO sightings over the nation's capitol. These sightings occurred from July 12, 1952 through July 27, 1952.

One of the most dramatic examples occurred on July 19, 1952, when an air traffic controller at Washington National Airport observed unknown objects on his radar. Edward Nugent counted seven objects at an estimated distance of 15 miles. Along with his supervisor and other controllers, Nugent checked his radar and they all concluded that it was operating properly. They contacted another control tower. The controllers in that tower confirmed the objects on radar and also stated that from their window they could see one of the objects in the shape of a bright orange light.

U.S. Capitol

To everyone's surprise, the object flew over the United States Capitol and the White House. Reports of other people seeing the objects poured in, including from an airman, persons in control towers, runways – and from a Capital Airlines pilot who was waiting in his airplane to take off. The pilot reported seeing six objects during fourteen minutes – all confirmed by the radar at a control tower.

The objects at times stood still while at other times made abrupt changes in position, altitude and speed. Two jets were scrambled from Newcastle AFB in Delaware. Just before the jets arrived, the objects disappeared – only to reappear after the jets ran low on fuel and had to return to their base. This sudden return convinced some witnesses that the objects were guided by intelligent beings.

Although these events caused serious concern on the part of President Truman and the public, the "Washington National Airport Sightings" or "Washington Flap," as they were soon called, produced typical unconvincing explanations from the Air Force. General John Samford and General Roger Ramey attributed the sightings to misidentified aerial phenomena and temperature inversion. Although their explanation would suffice if there was one incident, one witness and one control tower, it does not adequately address how multiple towers, multiple witnesses – all from different locations and over a span of weeks could see objects in the sky moving at incredible speeds and making impossible maneuvers.

Exhibit 25
Betty and Barney Hill

An American couple from the state of New Hampshire reported that on September 19, 1961 they experienced what we now refer to as an alien abduction.

Barney, a U.S. postal worker, and Betty, a social worker, were in New Hampshire and in their car driving back from a Canadian vacation, when around 10:30 p.m. they encountered a bright light in the sky. The light grew brighter and moved in different directions. It soon became clear to the Hills that they weren't observing any meteorite or airplane. At one point the disk-like object descended and hovered in front of their car, no more than one hundred feet away. The Hills reported seeing humanoid forms gazing out the craft's windows and Barney remembered telepathic communication with the aliens instructing him to stay where he was – just before both he and Betty lost consciousness.

When they regained full consciousness, they discovered that they had travelled 35 miles

Flying Saucer Photo

further down the road. They continued their trip home, confused as to what might have happened. During the next two days fragments of memories surfaced, prompting the Hills to report their UFO encounter to Pease Air Force Base. The investigation concluded that there wasn't sufficient data to establish an alien and UFO connection and that the flying object was probably the planet Jupiter or other optical condition.

After Betty experienced a series of frightening lucid dreams that included a disk shaped object, grey aliens that were about five feet tall, and a forced medical examination, they sought other help from UFO groups and hypnosis experts. In 1966 the book *Interrupted Journey* by John G. Fuller was published. It detailed Betty and Barney's encounter and their efforts to understand it.

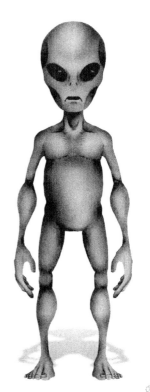

The public's reactions to the Betty and Barney Hill's alleged encounter range from complete dismissal of the account to full acceptance of it as a genuine alien and UFO experience. Regardless of varying conclusions, one thing is certain: the incident served to draw the attention of the public to the possibility that we are not alone in the universe – and to the possibility that we have visitors on Earth.

"Grey Alien" Drawing

Exhibit 26
Japan Air Lines

Japan Air Lines (1986 UFO Incident) occurred on November 17, 1986. The airplane, a Japanese cargo jumbo freighter, was on a flight from Paris to Tokyo, when at 5:11 p.m. near Alaska, two UFOs appeared on the left – close enough for the airplane's pilot to feel their heat on his face.

The two UFOs were soon joined by a third disk-shaped object, much larger, that soon started trailing them. The crew of the airplane tried to evade but were not successful. The entire event lasted 50 minutes.

The reporting of the incident resulted in the suspension of the captain of the cargo airplane, Captain Terauchi. He was grounded to a desk job and was only reinstated as a pilot years later.

A quote from the Wikipedia article, "Japan Air Lines flight 1628 incident" helps to appreciate the drama of the moment, as experienced by the pilot and his crew:

"As soon as JAL 1628 straightened out of its turn, at 17:11 PM, Captain Terauchi noticed two craft to his far left, and some 2,000 ft (610 m) below his altitude, which he assumed to be military aircraft. These were pacing his flight path and speed. At 5:18 or 5:19 PM the two objects abruptly veered to a position about 500 ft (150 m) or 1,000 ft (300 m) in front of the aircraft, assuming a stacked configuration.

In doing so they activated "a kind of reverse thrust, and [their] lights became dazzlingly bright". To match the speed of the aircraft from their sideways approach, the objects displayed what Terauchi described as a disregard for inertia: "The thing was flying as if there was no such thing as gravity. It sped up, then stopped, then flew at our speed, in our direction, so that to us it [appeared to be] standing still. The next instant it changed course. ... In other words, the flying object had overcome gravity." The "reverse thrust" caused a bright flare for 3 to 7 seconds, to the extent that captain Terauchi could feel the warmth of their glows."

Exhibit 27
Lights Over Phoenix

On March 13, 1997 residents of the Mexican state of **Sonora,** as well as residents of Arizona and Nevada, witnessed a series of lights in the sky that soon made national headlines.

The first lights were seen above Henderson, Nevada, at 18:55 PST. A man reported six lights that were attached to a huge, black V-shaped object. He described the object as being as big as a Boeing 747.

The object was later described by other witnesses as being solid, as it blocked the light of stars behind it.

An hour and twenty minutes later, at 20:15 MST, a police officer from Paulden, Arizona also reported seeing a group of lights.

Reports were also received from witnesses in Prescott and Prescott Valley. The witnesses reported that the object flew above their heads. Most of them commented that the craft didn't make any sound.

Sightings continued at the town of Dewey, 10 miles from Prescott, Arizona.

Soon thereafter lights were also seen in a V-shaped formation above the city of Phoenix.

Analysis of the sightings revealed that there were two separate events that night:

1. the lights in a V-shape craft that flew over 300 miles, and
2. the V-shaped lights reported over the city of Phoenix

The first sighting was of a moving V-shape. The second set of lights seemed to hover over the city of Phoenix.

The Air Force reported that the lights over the city of Phoenix were actually illumination flares dropped by the air force as part of a training mission. This explanation, even if 100 percent accurate, does not begin to explain at all the earlier V-shaped lights that flew across the state of Arizona and were seen and reported by many people.

Arizona Governor Fife Symington III stated, "It was enormous and inexplicable. Who knows where it came from? A lot of people saw it, and I saw it too. It was dramatic. And it couldn't have been flares because it was too symmetrical. It had a geometric outline, a constant shape."

Some observers state that the appearance of the lights over Phoenix (explained to be air force flares) was a very convenient and unusually coincidental occurrence, right after the sighting of the moving V-shaped object – which the air force never explained. When Governor Symington questioned the air force as to what the moving V-shaped lights were, the air force's response was "no comment."

Quote from Edgar Mitchell, U.S. Astronaut:

In 2004 he told the St. Petersburg Times that a "cabal of insiders" in the U.S. government were studying recovered alien bodies, and that this group had stopped briefing U.S. Presidents after John F. Kennedy.

He said, "We all know that UFOs are real; now the question is where they come from."

from Wikipedia article "Edgar Mitchell"

THE TESTIMONY OF EXPERTS

6

Astronauts: Expert Witnesses / 54

Carl Sagan and the Ancient Sumerian Language / 56

The Unknown Universe / 58

Exhibit 28
Astronauts: Expert Witnesses

In a court of law the testimony of witnesses is often used to prove a case. There is a class of witnesses referred to as "expert witnesses" whose testimony has more weight because of their expertise in a given field. For this reason, doctors are "expert witnesses' in a medical malpractice case, just as astronauts would be "expert witnesses" in a case involving extraterrestrial matters such as aliens and UFOs.

In doing research for this book, I read numerous accounts of pilots, astronauts, and other professionals in the aviation field and in space research who had seen UFOs and who held a strong and unshakeable belief in what they had witnessed. One of the many accounts that impressed me was that of astronaut Gordon Cooper, detailed in the Wikipedia article, "Gordon Cooper," a section of which is as follows:

"In 1957, when Cooper was 30 and a captain, he was assigned to Fighter Section of the Experimental Flight Test Engineering Division at Edwards Air Force Base in California. He acted as a test pilot and project manager. On May 3 of that year, he had a crew setting up an Askania-cinetheodolite precision landing system on a dry lake bed. This cinetheodolite system would take pictures at one frame per second as an aircraft landed. The crew consisted of James Bittick and Jack Gettys who began work at the site just before 0800, using both still and motion picture cameras. According to his accounts, later that morning they returned to report to Cooper that they saw a "strange-looking saucer" like aircraft that did not make a sound either on landing or take off.

"According to his accounts, Cooper realized that these men, who on a regular basis have seen experimental aircraft flying and landing around them as part of their job of filming those aircraft, were clearly worked up and unnerved. They explained how the saucer hovered over them, landed 50 yards away from them using three extended landing gears and then took off as they approached for a closer look. Being photographers with cameras in hand, they of course shot images with 35 mm and 4-by-5 still cameras as well as motion film. There was a special Pentagon number to call to report incidents like this. He called and it immediately went up the chain of command until he was instructed by a general to have the film developed (but to make no prints of it) and send it right away in a locked courier pouch. As he had not been instructed to not look at the negatives before sending them, he did. He said the quality of the photography was excellent as would be expected from the experienced photographers who took them. What he saw was exactly what they had described to him. He did not see the movie film before everything was sent away. He expected that there would be a follow up investigation since an aircraft of unknown origin had landed in a highly classified military installation, but nothing was ever said of the incident again. He was never able to track down what happened to those photos. He assumed that they ended up going to the Air Force's official UFO investigation, Project Blue Book, which was based at Wright-Patterson Air Force Base.

"He held claim until his death that the government is indeed covering up information about UFOs. He gives the example of President Harry Truman who said on April 4, 1950, "I can assure you that flying saucers, given that they exist, are not constructed by any power on Earth." He also pointed out that there were hundreds of reports made by his fellow pilots, many coming from military jet pilots sent to respond to radar or visual sightings from the ground. He was quite convinced till the day he died that he had seen UFOs and was a strong advocate to make the government come clean with what it knew.

"In his memoirs Cooper wrote he had seen other unexplained aircraft several times during his career and also said hundreds of similar reports had been made, often by military jet pilots responding to radar or visual sightings from the ground. He further claimed these sightings had been "swept under the rug" by the US government. Throughout his later life Cooper expressed repeatedly in interviews he had seen UFOs and described his recollections for the documentary Out of the Blue."

Exhibit 29
Carl Sagan and the Ancient Sumerian Language

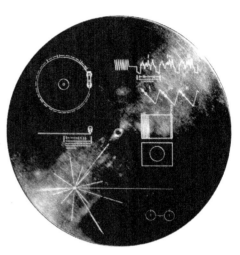

"Golden Record"

Carl Sagan (11/9/1934 – 12/20/1996) was a famous American cosmologist, astrophysicist, astronomer, professor and author. He taught at Harvard and Cornell and wrote popular books on astronomy and the evolution of life. He served as an advisor to NASA and wrote a science fiction novel "Contact" which was made into a film of the same name. Sagan's scientific insights included predicting atmospheric conditions on Venus, the possibility of liquid compounds on Titan, and water on Europa.

Sagan developed the "gold-anodized plaque" attached to Pioneer 10 (launched in 1972) and Pioneer 11 (launched in 1973). He also designed the content of the golden record attached to the Voyager space probes in 1977 which described our world and its inhabitants, with both written messages and oral greetings in a number of languages (See illustration above).

The golden record served as a friendly "message in a bottle", intended for any alien life form that might one day chance upon it.

The great science fiction author, Isaac Asimov, praised Carl Sagan by saying that he was one of only two people whose intellect was greater than his own.

Although Sagan advanced scientific inquiry regarding the possibility of intelligent aliens, he voiced skepticism of UFO reports – both historical and present day.

Then why are we mentioning him in this book whose subject is aliens and UFOs? The answer to the question lies in Sagan's curious selection of an obscure and dead language to record one of the greetings for his "message in a bottle." Out of 6,000 languages to choose from, Sagan selected 55 languages. One of them was the ancient (and now unused and quite dead) Sumerian language. Why choose this dead language?

Sumerian clay tablet

According to Zecharia Sitchin, a proponent of the ancient astronaut theory, the Anunnaki of Sumerian mythology are sentient creatures of a planet which comes close to the Earth every 3,600 years. Sitchin proposed that this cycle resulted in an interaction between the ancient Sumerians and the Anunnaki. Although this theory is widely debated and not generally accepted, it is very interesting that Carl Sagan, a scientist, a free thinker and skeptic, chose to include a Sumerian greeting in his "message in a bottle."

Exhibit 30
The Unkown Universe

Although some religions believe that everything was created only a few thousands of years ago, present day scientific consensus is that the age of the universe (everything that exists) is approximately 13.75 billion years old. Consensus is also that there are hundreds of billions of galaxies and perhaps trillions of worlds.

Scientists explain that the universe started with what they term a "big bang" - a mighty explosion that expanded a tiny bit of *something* to what is now *something else* (everything in the universe) and more than 93 billion light years wide! (and getting bigger every second).

Almost impossible to comprehend? Perhaps. Almost impossible to believe? Yes. (Makes my head spin.)

To make all this more incredible, some scientists hypothesize that in addition to the visible universe, there is much "dark matter' in the universe which no one - including the brightest scientists - have been able to see.

Some scientists also propose that we live in one universe out of many (multi-verses) and that the laws of physics in our universe don't necessarily apply in the others.

Is either of the following more credible than the other?

1. the belief that everything in the universe started from a tiny bit of something smaller than the period at the end of a sentence and that we are surrounded by "dark matter" which no one has been able to see,

or

2. among the trillions of worlds there might be life forms – some of which might be intelligent and have the ability to travel in space – and perhaps in time.

EXHIBIT 31: SELECT LIST OF UFOs IN HISTORY*

* The following list of Alien and UFO encounters was compiled from the following Wikipedia articles.

(For clarity, some parts have been edited or combined)

List of alleged UFO sightings

List of alleged aircraft-UFO incidents and near misses

UFO Sightings In the United States

Alien Abduction

Please note that this list is a selective list and is only a small sample of the numerous Alien and UFO reports that have been recorded since the dawn of man.

ALIENS AND UFOs: CASE CLOSED / 60

196 BC *angel hair*	Historian Cassius Dio (Roman) wrote "A fine rain resembling silver descended from a clear sky upon the Forum of Augustus." He used some of the material to plate some of his bronze coins, but by the fourth day afterwards the silvery coating was gone.
74 BC *flame-like "pithoi" from the sky*	According to Plutarch, a Roman army commanded by Lucullus was about to begin a battle with Mithridates VI of Pontus when "all on a sudden, the sky burst asunder, and a huge, flame-like body was seen to fall between the two armies." Plutarch reports the shape of the object as like a wine-jar (pithōi). The apparently silvery object was reported by both armies.
150 *100 foot "beast" accompanied by a maiden*	On a sunny day near the Via Campana, a road connecting Rome and Capua, a single witness, probably Hermas the brother of Pope Pius I, saw "a 'beast' like a piece of pottery (ceramos) about 100 feet in size, multicolored on top and shooting out fiery rays, landed in a dust cloud, accompanied by a "maiden" clad in white. Vision 4.1-3. in The Shepherd of Hermas.
10/24/1886 *Mystery Airships*	In a letter printed in the December 18, 1886 issue of *Scientific American*, page 389, the US consul of Venezuela in Maracaibo reported a UFO sighting.

	A bright object, accompanied with a humming noise, appeared during a thunderstorm over a hut near Maracaibo, causing its occupants to display symptoms similar to those from radiation poisoning. Nine days later the trees surrounding the hut withered and died.
1896-1897 *Aurora Texas UFO Incident*	Numerous reports of UFO sightings, attempted abductions that took place around the United States in a 2-year period.
1897	In a 1897 edition of the Stockton, California *Daily Mail*, Colonel H. G. Shaw claimed he and a friend were harassed by three tall, slender humanoids whose bodies were covered with a fine, downy hair who tried to kidnap the pair.
4/17/1897	A tale of a UFO crash and a burial of its alien pilot in the local cemetery was sent to newspapers in Dallas and Fort Worth (USA) in April 1897 by local correspondent S.E. Hayden.
6/30/1908 *Tunguska Event*	The explosion in Russia is believed to have been caused by the air burst of a large meteoroid or comet fragment at an altitude of 5--10 kilometres (3--6 mi) above the Earth's surface.

	Different studies have yielded varying estimates of the object's size, with general agreement that it was a few tens of metres across. The Tunguska event is the largest impact event over land in Earth's recent history. Impacts of similar size over remote ocean areas would most likely have gone unnoticed before the advent of global satellite monitoring in the 1960s and 1970s. UFO enthusiasts however class it as an exploding UFO.
1909 *Mystery Airships*	Strange moving lights and some solid bodies in the sky were seen around Otago and elsewhere in New Zealand, and were reported to newspapers.
8/13/1917 9/13/1917 10/13/1917 *Miracle of the Sun*	Thousands of people observed the sun gyrate and descend (Portugal). This was later reinterpreted by Jacques Vallée, Joaquim Fernandes and Fina d'Armada as a possible UFO sighting, but not recognized as such due to cultural differences.
1940s *Foo Fighters*	Small metallic spheres and colorful balls of light repeatedly spotted and occasionally photographed worldwide by bomber crews during World War II. (Europe and Asia)

ALIENS AND UFOs: CASE CLOSED / 63

1942	On 25 March, over the Zuiderzee an RAF Vickers Wellington from No. 301 Polish Bomber Squadron at RAF Hemswell, commanded by 2nd Lt Roman Sobinski was followed by an orange-coloured luminous disc on a journey back from a mission over Essen in the Ruhr Valley.
	The disc followed the bomber for five minutes at a distance of 150m and height of 14,000 ft and was fired on by the tail gunner, then disappeared at a very high speed.
	Such sightings were later christened *foo fighters* and many were seen in November 1944 on bombing missions over Germany.
1942 *Hopeh Incident*	A UFO was spotted and photographed (China).
2/24/1942 *Battle of Los Angeles*	Unidentified aerial objects trigger the firing of thousands of anti-aircraft rounds and raise the wartime alert status (in USA).
1946 *The Ghost Rockets*	Numerous UFO sightings were reported over Scandinavia; Swedish Defense Staff expressed concern.
5/18/1946 *UFO-Memorial Angelholm*	Gösta Karlsson (Sweeden) reports seeing a UFO and its alien passengers.
	A model of a flying saucer is now erected at the site.

ALIENS AND UFOs: CASE CLOSED / 64

6/21/1947 *Maury Island Incident*	Harold A. Dahl (United States) reported that his dog was killed and his son was injured by encounters with UFOs. He also claimed that a witness was threatened by the Men in Black.
6/24/1947 *Kenneth Arnold UFO Sighting*	The UFO sighting that sparked the name *flying saucers*. This (USA) sighting is considered as the start of the "Modern UFO era". Arnold said he saw nine unusual objects flying in a chain near Mount Rainier, Washington while he was searching for a missing military aircraft in his CallAir A-2. He described the objects as almost blindingly bright when they reflected the sun's rays; their flight as "erratic" ("like the tail of a Chinese kite") and flying at "tremendous speed." Arnold's story was widely carried by the Associated Press and other news outlets, and is usually credited as the catalyst for modern UFO interest, though many less-publicized UFO incidents preceded it.
6/1947 *UFO Sightings*	Several UFO sightings reported a few hours after the sighting of Kenneth Arnold.
7/8/1947 *Roswell UFO Crash*	United States Army Air Forces allegedly captures a crashed saucer and its alien occupants.

	The Roswell UFO Incident involved the recovery of materials near Roswell, New Mexico, in July 1947, which have since become the subject of intense speculation and research.
	There are widely divergent views on what actually happened, and passionate debate about what evidence can be believed.
	The United States military maintains that what was recovered was a top-secret research balloon that had crashed.
	However, many UFO proponents believe the wreckage was of a crashed alien craft and that the military covered up the craft's recovery.
	The incident has evolved into a recognized and referenced pop culture phenomenon, and for some, Roswell is synonymous with the term UFO, and likely ranks as the most famous alleged UFO incident.
1948 *The Green Fireballs*	Objects were reported over several United States military bases; an official investigation followed.
1/7/1948 *Thomas Mantell*	The Mantell Incident is among the most publicized early UFO reports: the crash and death of 25-year-old Kentucky Air National Guard pilot, Captain Thomas F. Mantell, on January 7, 1948, while in pursuit of a UFO.
	Historian David Michael Jacobs argues that the Mantell case marked a sharp shift in both public and governmental perceptions of UFOs.
	Previously, mass media often treated UFO

	reports with a whimsical or glib attitude reserved for silly season news.
Following Mantell's death, however, Jacobs notes "the fact that a person had dramatically died in an encounter with an alleged flying saucer dramatically increased public concern about the phenomenon.
Now a dramatic new prospect entered thought about UFO's: they might be not only extraterrestrial but potentially hostile as well."
7/24/1948
Chiles-Whitehead UFO Encounter	The Chiles-Whitted UFO Encounter is alleged to have occurred on July 24, 1948.
Two American commercial pilots reported that their Douglas DC-3 had nearly collided with a strange torpedo shaped object flying near them.
It was an important UFO sighting for several reasons:
It was perhaps the first that occurred at close distance (allegedly within a few hundred feet); and it was reported by two very experienced pilots, Clarence Chiles and John Whitted.
Both pilots had been decorated for their service as airmen during World War II, and both were regarded as valuable, respectable employees of Eastern Airlines. Chiles, in particular, was highly esteemed by his peers and by his employer.
It was a pivotal case for personnel of the US Air Force's Project Sign, and was a main reason they championed the extraterrestrial hypothesis as best explanation for UFOs.

ALIENS AND UFOs: CASE CLOSED

10/1/1948 *Gorman Dogfight*	A US Air Force pilot sighted and pursued a UFO for 27 minutes over Fargo, North Dakota.
1949 *Frank Scully*	An alleged retrieval of a grounded UFO and its occupants from a plateau in New Mexico.
4/1950 *Varese Close Encounter*	A factory worker (in Italy) sighted three Humanoids near a craft. One of the entities saw him and shot him with a beam, although it did not do any harm to the man.
5/1950 *Dr. Alexandro Botta UFO Incident*	A famous controversial case wherein Doctor A. Botta (United States) claimed to have entered a crashed UFO over a plateau.
1950 *Mariana UFO Incident*	The manager of Great Falls' (Montana, USA) pro baseball team took color film of two UFOs flying over Great Falls. The film was extensively analyzed by the US Air Force and several independent investigators.
5/11/1950 *McMinnville UFO Photographs*	Two farmers (Oregon, USA) took pictures of a purported "flying saucer." These are among the best known UFO pictures, and continue to be analyzed and debated to this day.

ALIENS AND UFOs: CASE CLOSED

8/25/1951 Lubbock Lights	Several Lights in V-Shaped formations were repeatedly spotted flying over the city (Lubbock, Texas). Witnesses included professors from Texas Tech University and photographed by a Texas Tech student.
1952 Operation Mainbrace	From 14 to 25 September many UFOs were observed. On 19 September at 11am a silver disc-shaped object followed a Gloster Meteor returning to RAF Topcliffe, and seen by observers on the ground. It rotated whilst hovering. It then travelled towards the west at high speed. On 21 September, six RAF planes followed a spherical object over the North Sea. It followed one of the planes back to the base.
7/13/1952 1952 Washington D.C. UFO Incident	A series of sightings in July 1952 accompanied radar contacts at three separate airports in the Washington area. The sightings made front page headlines around the nation, and ultimately lead to the formation of the Robertson Panel by the CIA.
7/24/1952 Carson Sink UFO Incident	The Carson Sink Case was a famous UFO incident that occurred over the Carson Sink in western Nevada in the United States on July 24, 1952. The incident is considered especially

	noteworthy among UFO sightings because of the competency and reliability of the witnesses, two experienced command pilots of the United States Air Force.
9/12/1952 *The Flatwoods Monster*	Six local boys and a woman (Flatwoods, West Virginia) report seeing a UFO land, and saw a spade-headed creature near the landing site.
5/21/1953 *Prescott Sightings*	Three Prescott residents (Prescott, Arizona) sighted a total of eight craft at Del Rio Springs Creek, 20 miles north of Prescott.
8/12/1953 *Ellsworth UFO Case*	A UFO appearing as a red glowing light (Bismarck, North Dakota) is witnessed by 45 people. The sighting takes place over a two night period.
11/23/1953 *Disappearance of Felix Moncla*	Felix Moncla, Jr. was a United States Air Force pilot who disappeared with 2nd Lt. Robert Wilson while pursuing an unidentified flying object over Lake Superior in 1953. The US Air Force reported that Moncla had crashed and that the "unknown" object was a misidentified Canadian Air Force airplane, but the Royal Canadian Air Force (RCAF) disputed this claim, reporting that none of their craft were near the area in question.

ALIENS AND UFOs: CASE CLOSED / 70

12/16/1953 *Kelly Johnson/Santa Barbara Channel Case*	Legendary Lockheed aircraft engineer Clarence "Kelly" Johnson, designer of the F-104, U-2, and SR-71, and his wife observed a huge Flying Wing over the Pacific from the ground in Agoura. Meanwhile, one of Johnson's flight test crews aboard an WV-2 (see EC-121) spotted the craft from Long Beach, California. USAF concluded these trained observers had seen a lenticular cloud, even though Johnson considered and ruled out that explanation.
1954	On 14 October Flt Lt James Salandin of the Royal Auxiliary Air Force, flying in a No. 604 Squadron RAF Gloster Meteor F8 from RAF North Weald, narrowly missed two UFOs over Southend-on-Sea at around 4.30pm at 16,000 ft. The objects were circular with one being colored silver and the other gold. He narrowly avoided having a head-on collision with the silver object.
8/21/1955 *Kelly-Hopkinsville Encounter*	A group of strange, goblin-like creatures are reported to have attempted to attack a farm house (Kentucky, USA). The family shot at them several times with little or no effect.

ALIENS AND UFOs: CASE CLOSED

1956 *Lakenheath-Bentwaters Incident*	Lakenheath-Bentwaters incident – on 13 August, 12 to 15 objects were picked up by USAF radar over East Anglia. One object was tracked at more than 4,000 mph by USAF GCA radar at RAF Bentwaters. The objects sometimes travelled in formation, then converged to form a larger object and performed sharp turns. One object was tracked for 26 miles which then hovered for five minutes then flew off. One object at 10pm was tracked at 12,000 mph. RAF de Havilland Venoms from RAF Waterbeach had sightings of the objects.
1957 *Antonia Villas Boas*	Antonio Villas Boas claimed to have been abducted and examined by aliens (Sao Francisco de Sales, Brazil). He also claimed to have had sex with an alien woman while aboard the UFO.
1957 *West Freugh Incident*	On 4 April, a large object was seen on radar at RAF West Freugh near Stranraer at 50,000 ft which was stationary for 10 minutes over the Irish Sea. It moved vertically to 70,000 ft and was also tracked by radar at Ardwell. The object did an 'impossible' sharp turn and was described as being as large as a ship, bigger than a normal aircraft.

ALIENS AND UFOs: CASE CLOSED / 72

5/3/1957 Edwards Air Force Base UFO	Jack Gettys and James Bittick, who were filming base installations on behalf of Gordon Cooper, (California, USA) observed the landing and departure of a flying disk. Their film evidence was sent to Washington.
5/20/1957 Milton Torres 1957 UFO Encounter	US Air Force fighter pilot Milton Torres reports that he was ordered to intercept and fire on a UFO displaying "very unusual flight patterns" over East Anglia (England). Ground radar operators had tracked the object for some time before Torres' plane was scrambled to intercept.
9/1957 Ubatuba UFO Explosion	Two Fishermen watched a UFO crash and explosion, and retrieved fragments of the object (Ubatuba, Brazil).
10/1957 Antonio Villas Abduction	One of the very first abduction claims (Sao Francisco de Sales, Brazil)
11/2/1957 Levelland UFO Case	Numerous people describe seeing a glowing, egg-shaped object and a cigar-shaped object which caused their vehicle's engines to shut down. The Levelland UFO Case occurred on November 2–3, 1957, in the small town of Levelland, Texas. Levelland, which in 1957 had a population of about 10,000, is located west of Lubbock on the flat prairie of the Texas

	panhandle. The case is considered to be one of the most impressive in UFO history, mainly because of the large number of witnesses involved over a relatively short period of time.
11/4/1957 *Fort Itaipu UFO Invasion*	UFO attacks the sentinels of military unit. (Praia Grande, Sao Paulo, Brazil)
12/1957 *Old-Saybrook UFO Incident*	Three separate sightings possibly describing the same cigar-shaped object, all of which sighted the occupants inside. (Connecticut, Unites States)
1/16/1958 *Trinidade UFO Incident*	Nine separate sightings and 7 photos of UFO's were reported in the Trinidade Island (Brazil) during the meteorological and geological expeditions in the island.
1959 *Dyatlov Pass Incident*	Mysterious deaths of experienced skiers in the Urals are believed to have been caused by "unidentified orange spheres" and an "unknown compelling force". (Russia)
2/24/1959	An American Airlines Douglas DC-6 was followed for 45 minutes by three saucer shaped objects on its flight from Newark to Detroit. It was seen by the crew and the 35 passengers, and by Captain Killian.

ALIENS AND UFOs: CASE CLOSED / 74

6/26/1959 – 6/27/1959 *Father William Booth Gil Sighting*	Missionary and 25 independent witnesses saw a UFO being repaired by 4 human-like occupants. (Boionai, Papua New Guinea)
9/19/1961 *Betty and Barney Hill Abduction*	The first widely publicized alien abduction experience. (New Hampshire, United States)
4/24/1964 *Lonnie Zamora*	Police officer Zamora reports a close encounter.
9/4/1964 *Donald Shrum*	The hunter got lost in a forest and took shelter in a tall tree. After flashing a flashlight in the air, a UFO approached the tree and two Humanoids and one Robot repeatedly attempted to climb the tree and abduct him, he stopped the entities by shooting them with arrows. (Cisco Grove, California)
3/18/1965	A Convair CV-240 from Toa Airways heading from Osaka to Hirshima was followed over Lejima by an oblong 15 metre long green luminescent object, forcing the pilot, Yoshiharu Inaba, to make a 60 degree turn to avoid a collision. The object stopped and flew alongside the plane for about three minutes, and affected the Automatic Direction Finder.

ALIENS AND UFOs: CASE CLOSED / 75

3/9/1965 *The Incident At Exeter*	The Exeter incident of Exeter, New Hampshire occurred on September 3, 1965. A UFO the size of a barn was seen as close as 500 feet away by a teenager and two police officers.
12/9/1965 *Kecksburg UFO Incident*	Mass sighting of a falling UFO, followed by a cordoning-off of the crash site by alleged US military personnel.
1966 *The Mothman Prophecies*	A wave of reported sightings of a winged humanoid are connected to other mysterious events, including sightings of UFOs.
3/1966 *Michigan Swamp Gas Sightings*	Widely reported wave of sightings with a large number of law enforcement witnesses. Originally attributed to "swamp gas" by J. Allen Hynek. Watershed case that brought a spotlight of public doubt on "official" UFO investigations, instigated a congressional inquiry by then-Representative Gerald R. Ford. The sighting appears to be the "turning point" for Hynek. (Michigan, USA)
4/6/1966 *Westfall UFO*	A sighting reported by hundreds of people (Clayton South, Victoria, Australia). Witnesses of "The Clayton Incident" still gather for reunions.

ALIENS AND UFOs: CASE CLOSED / 76

4/17/1966 *Portage County UFO Chase*	Several police officers pursue what they believe to be a UFO for 30 minutes. (Ohio, United States). The so-called Portage County UFO Chase was an unidentified flying object encounter that began in Portage County, Ohio on the morning of April 17, 1966, when police officers Dale Spaur and Wilbur Neff observed a metallic, disc shaped object flying in the skies. They pursued the object for about half an hour, ending up in Pennsylvania before losing sight of the UFO. Several other police officers became involved in the chase, and several civilians reported witnessing the same, or a similar object in about the same area, during this time.
10/11/1966 *The Grinning Man*	A tall man with no nose or ears is reported in a neighborhood shortly after UFO sighting. (Elizabeth, New Jersey)
1966 *Indrid Cold*	The sightings of the "Grinning Man" in Point Pleasant. (West Virginia, United States)
1/25/67 *Betty Andreasson Abduction*	An alleged abduction of eleven people by pear-shaped-headed creatures from a red UFO. (South Ashburnham, Massachusetts, US)

ALIENS AND UFOs: CASE CLOSED / 77

5/20/1967 *Falcon Lake Incident*	A man is reported to have been burned by the exhaust from a UFO. (Falcon Lake, Manitoba)
8/29/1967 *Close Encounter of Cassac*	A young brother and sister claim to have witnessed a UFO and its occupants. (Cussac, Cantal, France)
9/1/1967 *Snippy the Horse Mutilation*	Widely considered to be the first unusual animal death to be related by its witnesses to UFOs and aliens. (San Luis Valley, Colorado, USA)
10/4/1967 *Shag Harbour UFO Crash*	A UFO was reported to have crashed into Shag Harbor. A Canadian naval search followed, and officially referred to the incident as a UFO crash. (Nova Scotia, Canada) Nebraska Police Sergeant Herbert Schirmer claimed that he was abducted by extraterrestrials in 1967. His case was one of those investigated in the Condon Report. He flew to Boulder, Colorado and was examined under hypnosis by psychologist Dr. R. Leo Sprinkle of the University of Wyoming on February 13, 1968.

ALIENS AND UFOs: CASE CLOSED / 78

12/3/1967 *Herbert Schirmer*	Sergeant Herbert Schirmer claimed that he was abducted. (Ashland, Nebraska, USA)
8/1968 *Buff Ledge Camp Abduction*	Two teenagers reported a sighting of a UFO over Lake Champlain, and claimed to have experienced missing time.
1969 *Jimmy Carter UFO Incident*	Jimmy Carter's sighting. (Leary, Georgia) Jimmy Carter claimed that he had seen a UFO in 1969. In 1973, while Governor of Georgia, he filed a report with the International UFO Bureau in Oklahoma City, Oklahoma.
1/1/1970 *Cowichan District Hospital UFO Incident*	Several nurses and patients reported sighting a UFO with masked occupants. (Cowichan, Canada)
4/24/70	On April 24, a Soviet bomber disappeared on its flight from Moscow to Vladivostok without a trace. On the same day several UFOs were observed in the area of the Soviet-Chinese border, which could not be shot down by the Russian military.
5/1973 *The Judy Doraty Abduction*	Four people claimed to have been abducted from their car by aliens with egg-shaped heads, and to have witnessed a cattle mutilation. (Texas, USA)

ALIENS AND UFOs: CASE CLOSED

10/1973 *The Alabama Close Encounter*	A policeman describes seeing an alien which he described as being created out of "Tin Foil" (Alabama, USA)
10/11/1973 *Pascagoula Abduction*	Two men fishing on the Pascagoula River claimed to be abducted by strange looking humanoids. (Mississippi, USA)
10/17/1973 *Eglin Air Force Base Sighting*	An unidentified object was tracked by a Duke Field radar unit during the same time period, and within the same area, that 10 to 15 people observed four strange objects flying in formation between Milton, Florida, and Crestview, Florida, along Interstate 10, according to Eglin officials. Reports from the base indicated that a bright glowing ball of light could be seen travelling parallel with an Air Force C-130 aircraft but at a much higher altitude. (Florida, USA)
10/30/1975 *Wurtsmith AFB*	USAF security personnel reported an unidentified craft flying within exceptionally-secure Strategic Air Command airspace over a B-52 base housing nuclear weapons and delivery systems. An incoming KC-135 tanker was later ordered to commence pursuit over Lake Huron. The object or objects, last seen back over the base's weapons storage area, were never identified. (near Oscoda, Michigan, USA)

ALIENS AND UFOs: CASE CLOSED

11/5/1975 *Travis Walton*	Logger Travis Walton reports being abducted by aliens for five days. Walton's six workmates claimed to have witnessed the UFO at the start of his abduction. Walton described the event and its aftermath in the book *The Walton Experience*, which was dramatized in the film *Fire in the Sky*. (Arizona USA)
6/22/1976 *1976 Canary Isles Sightings*	Several lights and a spherical transparent blue craft, piloted by two beings was reported. (Canary Islands, Spain)
8/1976 *Stanford Abduction*	3 women were alleged to be abducted and harassed by aliens. (Kentucky, USA)
8/20/1976 *Allagash Abductions*	The Allagash Abduction is an incident that occurred on August 20, 1976, when four men, Jim Weiner, his twin brother Jack, Chuck Rak and their guide, Charlie Foltz, all in their early-twenties, ventured on a camping trip into the Allagash wilderness of Maine.
9/19/1976 *1976 Tehran UFO Incident*	A UFO disabled the electronic equipment of two F-4 interceptor aircraft, along with ground control equipment, an event thoroughly documented in the U.S. DIA report.

ALIENS AND UFOs: CASE CLOSED / 81

	The Iranian generals involved in the incident claimed the object was extraterrestrial. (Tehran, Iran)
1977 *Colares UFO Flap*	A bewildering account of an island attacked by UFOs shooting harmful beams of radioactive light at the residents. (Colares, Brazil)
5/10/1978 *Emilcin Abduction*	A man in Emilcin, Poland is said to have been abducted by "grays." There is now a memorial at the site.
10/21/1978 *Valentich Disappearance*	Contacting air traffic control, an Australian pilot reported seeing a UFO before his aircraft vanished. (Victoria, Australia)
12/21/1978 *Kaikoura Lights*	A series of sightings by a Safe Air freight plane; the airplane was escorted by strange lights that changed color and size. (South Island, New Zealand)
8/27/1979 *Val Johnson Incident*	A deputy sheriff spotted a bright light which appeared to have collided with his patrol car and damaged it. The deputy also suffered temporary retinal damage from the "light". (Marshall County, Minnesota, USA)

ALIENS AND UFOs: CASE CLOSED / 82

11/9/1979 *Dechmont Woods Encounter*	A forester, Bob Taylor, was pulled by two spiked globes towards a UFO, which stood on a clearing. He lost consciousness and afterwards had trouble walking and speaking. He was also constantly thirsty for several days. (Livingston, Scotland)
11/11/1979 *Manises UFO Incident*	Three large UFOs forced a commercial flight to make an emergency landing at Manises Airport. (Valencia, Spain)
12/28/1980 *Rendlesham-Woodbridge Incident*	A sighting by military personnel, which at first appeared to be a downed aircraft. (Suffolk, England)
12/29/1980 *Cash-Landrum Incident*	(New Caney, Texas, USA) A huge diamond-shaped UFO irradiates three witnesses, who all required treatment for radiation poisoning. The UFO was escorted by military helicopters. The victims have since sued the US Government. It might be classified as a Close Encounter of the Second Kind, due to its reported physical effects on the witnesses and their automobile.

ALIENS AND UFOs: CASE CLOSED / 83

1/8/1981 *Trans-en-Provence Case*	Renato Nicolai, a farmer, saw an object which had the shape of two saucers, one inverted on top of the other. The UFO left physical evidence on the ground, in the form of mechanical pressure and burnt residue on the grass. (Trans-en-Provence, France)
1981 *Hudson Valley Sightings*	A Wave of UFO cases in the Hudson Valley. (USA)
6/1983 *Copely Woods Encounter*	Hundreds of Basketball-Sized balls of light were sighted around a neighborhood, leaving unusually obvious marks behind. (Indiana, USA)
1985 *Whitley Strieber*	The abduction that took place in Whitley Strieber's Apartment. (New York, USA)
5/19/1986 *Sao Paolo UFO Sighting*	Brazilian Air Force detected and intercepted UFOs on southeastern Brazil. As many as twenty UFOs were seen and tracked by ground radar and at least six airplanes during the night of May 19. (Sao Paolo and Rio de Janeiro, Brazil)
11/17/1986 *Japan Air Lines flight 1628 Incident*	A group of UFOs flew alongside Japan Air Lines Flight 1628 for 50 minutes above Northeastern Alaska. One of the objects trailing the Boeing 747 was detected by military radar.

ALIENS AND UFOs: CASE CLOSED / 84

3/30/1990 *Belgian UFO Wave*	Mass sighting of large, silent, low-flying black triangles, which were tracked by multiple NATO radar and jet interceptors, and investigated by Belgium's military. Photographic evidence exists. (Ans, Wallonia, Belgium)
9/15/1991 STS-48 Incident	Video taken during mission STS-48 shows a flash of light and several objects, apparently flying in an artificial or controlled fashion. NASA explained them as ice particles reacting to engine jets. Philip C. Plait, in his book *Bad Astronomy*, agreed with NASA, but Jack Kasher proposed five arguments against them being ice particles. James Oberg disputed these, and Lan Fleming argued that the shuttle's exhaust plume as not the cause of the flash of light that preceded the objects' abrupt change of course. Mark J. Carlotto noted that one of the objects apparently had three lobes arranged in a triangular pattern. (Space Shuttle Discovery while in orbit)
3/26/1993 Kelly Cahill	Eight witnesses described seeing a large glowing UFO shortly after midnight. They also encountered 7 feet (2.1 m), void-colored aliens with glowing red eyes. (Dandenong, Australia)

1/28/1994	On 28 January on Air France Flight AF3532 in an Airbus A320 flying from Nice to London, Jean-Charles Duboc saw an unusual object near Paris at around 1pm which looked like a huge red-brown disc at 35,000 ft of approximately 800 ft in diameter. The object was reported to air traffic control at Reims in Champagne-Ardenne, and was determined to be close to Taverny Air Base, the headquarters of the French Air Force. It was stationary in the sky for around one minute then disappeared in around 10–20 seconds.
6/1994 Meng Zhaoguo Incident	Meng claimed to have been abducted and forced to have sexual intercourse with a 10 feet (3.0 m), six fingered, female alien with braided leg fur. (Wuchang, China)
5/25/1995 America West Airline Flight 564	A 300–400 foot long cigar-shaped UFO with rotating strobe light followed an America West Boeing 757. (Bovina, Texas)
1/20/1996 Varginha UFO Incident	Multiple sightings and the alleged capture of an alien by the Brazilian military. (Varginha, Minas Gerais, Brazil)
10/5/1996 Westendorff UFO Sighting	Pilot observes a UFO emerge from a mother craft. (Pelotas, Brazil)

ALIENS AND UFOs: CASE CLOSED

3/13/1997 Phoenix Lights	Lights and craft of varying descriptions, most notably a V-shaped pattern, were seen by thousands of people between 19:30 and 22:30 MST, in a space of about 300 miles, from the Nevada line, through Phoenix, to the edge of Tucson.
1/5/2000 Black Triangle (UFO)	The "St. Clair Triangle", "UFO Over Illinois", "Southern Illinois UFO", or "Highland, Illinois UFO" sighting occurred on January 5, 2000. It occurred over the towns of Highland, Dupo, Lebanon, Summerfield, Millstadt, and O'Fallon, Illinois, beginning shortly after 4:00 am. Five on-duty Illinois police officers in separate locales, along with various other witnesses, sighted and reported a massive, silent, triangular aircraft operating at an unusual range of near-hover to incredible high speed at treetop altitudes. The incident was examined in an ABC Special "Seeing is Believing" by Peter Jennings, an hour-long special "UFOs Over Illinois", produced by Discovery Channel, a Sci Fi Channel special entitled "Proof Positive" as well as a 28 minute independent documentary titled "The Edge of Reality: Illinois UFO, January 5, 2000" by Darryl Barker Productions, St. Louis, Missouri.
7/15/2001 NJ Turnpike/Carteret Lights Incident	At least 15 people, including 2 police officers, stopped their cars along the New Jersey Turnpike to view light formations in the night sky. (Carteret, New Jersey, USA)

10/31/2004 *The Tinley Park Lights*	A sequence of five mass UFO sightings, first on August 21, 2004, two months later on October 31, 2004, again on October 1 of 2005, and once again on October 31, 2006, in Tinley Park and Oak Park, Chicago. A triangular formation of reddish lights were seen at low to intermediate altitude by hundreds of witnesses, on three separate occasions in late 2004 and early 2005, producing multiple videos, photos, and mainstream local news coverage over two suburbs of Chicago, Illinois. The object(s) maneuvered slowly within a busy airspace near O'Hare International Airport. The incident was investigated by MUFON, and reported widely in metropolitan media.
11/7/2006 *Chicago O'Hare UFO Sighting 2006*	United Airlines employees and pilots claimed sightings of a saucer-shaped, unlit craft hovering over a Chicago O'Hare Airport terminal, before shooting up vertically. (Chicago, Illinois, USA)
2007 *Charles Hall*	Charles Hall claimed to have met and befriended several tall, pasty-white extraterrestrials while working on the Nellis Air Force Base. (Nevada, USA)
4/23/2007 *2007 Alderney UFO Sighting*	Two airline pilots on separate flights spot UFOs off the coast of Alderney. (Bailiwick of Guernsey, Crown Dependency)

ALIENS AND UFOs: CASE CLOSED / 88

9/25/2007 *Kodiak Island UFO Incident*	On September 25, 2007, several Kodiakans saw something fall from the sky on a Tuesday morning that may have landed in mountainous terrain on Kodiak Island. The incident prompted 911 calls and a helicopter search was launched from U.S. Coast Guard Air Station Kodiak, but no crash site was found. (Kodiak, Alaska, USA)
1/8/2008	On January 8, 2008, more than 30 residents of Stephenville, Texas reported seeing a large object described as being about a mile long and a half-mile wide with bright lights being chased by what appeared to be fighter jets. The Air Force claims they had no aircraft in the area at the time of the sightings and says the objects may have been an illusion caused by two commercial airplanes. One resident described the object as "an arch shape converted in a vertical shape, and then it split and made two of them, and then these turned into just fire and it was gone."
5/2008 – 9/2008 *2008 Turkey UFO Sightings*	Over a four month span in 2008, a night guard at the Yeni Kent Compound videotaped one or more UFOs over Turkey at nighttime. Many witnesses confirmed the two and a half hours' worth of video, leading the Sirius UFO Space Science Research Center to

ALIENS AND UFOs: CASE CLOSED

	dub it the "most important images of a UFO ever filmed." (Istanbul, Turkey)
6/20/2008 *Wales UFO Sightings*	According to media reports, a police helicopter was almost hit by a UFO, before it tried to pursue it. Hundreds of people reported to have witnessed a UFO on the same or preceding days, from different areas of Wales. (Different Cities, Wales, United Kingdom)
6/21/2008 *Moscow UFO Sightings*	Different people and media (including state-owned) reported sightings of 11 orange UFOs. Further confirmation came from Saint Petersburg and Novosibirsk. On June 25 a similar report, now with 13 objects, was claimed from the UK. (Moscow, Russian Federation)
1/25/2010 *Harbour Mille Incident*	At least three UFO's were spotted over Harbour Mille. The objects looked like missiles but emitted no noise. (Harbour Mille, Newfoundland and Labrador, Canada)

ALIENS AND UFOs: CASE CLOSED

2/20/2011 *Vancouver Washington UFO Sighting*	Stationary green and red blinking lights with limited sideways movements, were recorded on video and still photography. They were witnessed by several residents who live off Southeast 192nd Avenue in Vancouver, Washington.

    In a 1988 interview on Larry King's radio show, (Barry) Goldwater was asked if he thought the U.S. Government was withholding UFO evidence; he replied "Yes, I do."

    He added: "I certainly believe in aliens in space. They may not look like us, but I have very strong feelings that they have advanced beyond our mental capabilities....I think some highly secret government UFO investigations are going on that we don't know about – and probably never will unless the Air Force discloses them."

from the Wikipedia article "Barry Goldwater"

TO YOU: THE JUDGE AND JURY   8

In Section 1, the introduction of this book, we discussed evidence and the different levels of proof. We proposed introducing evidence that is competent, relevant and material and that more than suffices to prove the case for the existence of Aliens and UFOs under the strictest evidentiary standard, "proof beyond a reasonable doubt." Throughout the book we provided numerous examples in the form of exhibits.

Section 2, "Ancient Historical Perspective," includes selective examples of Alien and UFO sightings by our ancient ancestors – from cave dwellers to Christopher Columbus. They establish the fact that Alien and UFO sightings and encounters have been documented since the dawn of man. Much of the references to Aliens and UFOs are depicted or described in the limited terms familiar to the witnesses of the time and often explained within a restrictive religious framework.

Section 3, "Application Of Scientific Reasoning," presents scientific testimony for the existence of countless alien worlds and the almost certainty of intelligent life. It introduces man's attempt to scientifically classify and understand encounters while at the same coming to the realization that our understanding may be limited and elementary.

Section 4, "Common Sense Argument," discusses the relatively short span of man's existence, as compared to the probable longer span of older civilizations. It argues that we should be open to accepting the reality that we may not be the center of the universe and that we may still have much to learn.

Section 5, "Recent Historical Perspective," cites examples of recent Alien and UFO encounters. Radar, visual sightings, physical evidence and testimony of direct alien contact emphasize the fact that sightings and encounters are continuing and are becoming increasingly well documented.

Section 6, "The Testimony Of Experts," provides testimony – indirect and circumstantial (as in the case of Astronomer Carl Sagan) and very direct (as in the case of Astronaut Gordon Cooper). Expert witnesses are much sought out to testify in legal cases. In cases involving technical or scientific expertise their testimony is prized and weighed more heavily than that of a non-expert witness.

Section 7, "Select List Of UFOs In History," contains numerous descriptions of UFO and Alien sightings. This is a "select list" and would be unbelievably longer if it included all published reports. Although there may be some cases where the sighting may be attributable to error or fraud, we do not think that all can be summarily explained away with this reasoning.

The Judge and jury should keep in mind that the amount of available evidence in the form of testimony (from regular people and expert witnesses, art work, documents, films and investigative reports) is tremendous and that we have only provided in this book a small but sufficient enough sample to prove our case.

Based on the above, I ask that you act as the Judge and the Jury carefully consider the mountain of evidence that has been provided and agree with me that we have established our case that-

YES – Aliens and UFOs are real!

ALIENS AND UFOs: CASE CLOSED / 93

Illustrations

The illustrations in this book are either in the public domain, as credited below, or licensed from *Fotolia.com*.

Page	Description	Source
--	Cover (front and back)	*Fotolia.com*
1	Artist's rendering of Flying Saucer	*Fotolia.com*
2	Giordano Bruno	*Wikipedia.com (See note 1, below)*
5	Liberty holding scales of Justice	*Fotolia.com*
6	USA Supreme Court Building	*Fotolia.com*
7	Law Book	*Fotolia.com*
9	Ancient Cave Drawings	*Wikipedia.com (See note 2, below)*
13	The Great Pyramid at Giza, Egypt	*Fotolia.com*
13	The Pyramid of the Sun at Teotihuacan, Mexico	*Fotolia.com*
15	Monkey Line Drawing (Nazca Lines)	*Wikipedia.com. Released into the public domain by its author, Maria Reiche.*
17	Christopher Columbus	*Wikipedia.com (See note 2, below)*
18	Drawing from a Woodcut	*Wikipedia.com (See note 2, below)*
19	Religious Books	*Fotolia.com*
23	Statue of Giordano Bruno	*Wikipedia.com (See note 3, below)*

ALIENS AND UFOs: CASE CLOSED / 94

26	NASA's Kepler Telescope	*Wikipedia.com (See note 4, below)*
27	Extremophile	*Wikipea.com. (See note 5, below*
28	"Goldy Locks" Zones	*Wikipedia.com. (See note 4, below)*
29	First Five Kepler Planets	*Wikipedia.com (See note 4, below)*
30	DNA Double Helix Structure	*This work has been released into the public domain by its author, Thorwald, at the wikipedia project. This applies worldwide.*
31	Mars Meteorite ALH84001	*Wikipedia.com (See note 4, below)*
32	The Four Elements	*Wikipedia.com. This work has been released into the public domain by its author, Heron, at the English Wikipedia project. This applies worldwide.*
33	The Periodic Table Of Elements	*Wikipedia.com. (See note 3, below)*
37	Baby	*Fotolia.com*
39	Artist's Rendering Of An Alien City	*Fotolia.com*
41	Plaque At Aurora Cemetery	*Wikipedia.com. This work has been released into the public domain by its author, Sf46.*
43	Ball Lightning	*Wikipedia.com (See note 2, below)*
44	Roswell "Daily Record" Front Page	*Wikipedia.com (See note 7, below)*
46	U.S. Capitol	*Wikipedia.com (See note 8, below)*

48	Flying Saucer Photo	Wikipedia.com (See note 10, below)
49	"Grey Alien" Drawing	Wikipedia.com (See note 9, below)
56	"Golden Record"	Wikipedia.com (See note 4, below)
57	Sumerian Clay Tablet	Wikipedia.com (See note 2, below)

(1) This work is in the public domain in the United States, and those countries with a copyright term of life of the author plus 100 years or less.

(2) This work is in the public domain in the United States because its copyright has expired.

(3) This work is in the public domain because the copyright holder of the work has released it into the public domain.

(4) This file is in the public domain because it was solely created by NASA. NASA copyright policy states that "NASA material is not protected by copyright unless noted."

(5) This image or media file contains material based on a work of a National Park Service employee, created during the course of the person's official duties. As a work of the U.S. federal government, such work is in the public domain.

(6) This image is in the public domain because its copyright has expired. This work is in the public domain in the United States because it was published (or registered with the U.S. Copyright Office) before January 1, 1923.

(7) This work is in the public domain because it was published in the United States before 1964, and its copyright was not renewed.

(8) This image is a work of an employee of the Architect of the Capitol, taken or made during the course of the person's official duties. As a work of the U.S. federal government, all images created or made by the Architect of the Capitol are in the public domain, with the exception of classified information.

ALIENS AND UFOs: CASE CLOSED / 96

(9) Created by Wikipedia contributor LeCire and released by creator into the public domain.

(10) This work is in the public domain in that it was published in the United States between 1923 and 1977 and without a copyright notice.

Facts have been checked on Wikipedia.

A friendly alien from the children's book
"Pinocchio and the Dragons of Martoon"
by Angelo Tropea

ALIENS AND UFOs: CASE CLOSED / 97

A

aerial phenomena, 43, 47
aerial projects, 42
Aether, 32
Africa, 12
African, 10, 32
Air Force, 34, 35, 44, 47, 49, 51, 54, 55, 66, 67, 68, 70, 72, 79, 83, 85, 87, 88
aircraft, 42, 43, 44, 50, 54, 55, 59, 63, 64, 70, 71, 79, 80, 81, 82, 86, 88
airplane, 15, 21, 43, 47, 48, 50, 69, 81
airplane runways, 15
Alabama Close Encounter, 79
Alaska, 50, 83, 88
algae, 38
alien, 7, 10, 21, 26, 37, 42, 45, 48, 49, 56, 61, 63, 64, 71, 74, 79, 85, 91, 96
Alien contact, 37
Alien researchers, 10
aliens, 9, 10, 14, 20, 21, 31, 34, 39, 42, 45, 48, 49, 54, 56, 57, 71, 77, 78, 80, 84
Aliens, 3, 6, 7, 8, 9, 10, 31, 33, 34, 91, 92
Allagash Abductions, 80
Allied airplanes, 43
Allied forces, 14
America West Airline Flight 564, 85
ancestors, 9, 32, 37, 91
ancient aliens theory, 10
ancient astronaut theory, 57
ancient humans, 12
Annunaki, 57
anthropomorphic dummies, 45
Antonia Villas Boas, 71
Antonio Villas, 71, 72
archeologists, 38
Area 51, 42
Arizona, 51, 52, 69, 80
Asia, 12, 62
Asian, 10, 32
asteroids, 30
astronauts, 6, 10, 54
astronomy, 56
atmosphere, 38
atmospheric pressure, 28
atoms, 33
Aurora, 3, 40, 41, 42, 61, 94

Aurora Cemetery, 41
Aurora legend, 42
Aztecs, 12

B

bacteria, 30, 31
Battle of Los Angeles, 63
Betty and Barney Hill, 3, 40, 48, 49, 74
Betty Andreasson Abduction, 76
beyond a reasonable doubt, 1, 6, 91
Bhima, 16
big bang, 58
Book of Enoch, 19
Book of Ezekiel, 19, 20
Book of Genesis, 19
Brazil, 71, 72, 73, 81, 83, 85, 86
British Government's Ministry of Defense, 35

C

Canary Isles Sightings, 80
Capital Airlines, 46
Captain Terauchi, 50
Cargo Cults, 3, 8, 14
Carl Sagan, 3, 10, 53, 56, 57, 92, 105
Cash-Landrum Incident, 82
Cassius Dio, 60
cave dweller, 37
cave paintings, 21
caves, 37, 38
chariot, 16
Charles Hall, 87
Chicago O'Hare Airport, 87
Chiles-Whitehead, 66
Christian faith, 17
Christopher Columbus, 3, 8, 17, 91, 93
Church, 23
City of the Gods, 12
Civil Courts, 5
civilization, 9, 24, 35
civilizations, 12, 24, 31, 39, 91
classical times, 32
classification scales, 34
Clayton Incident, 75
Close Encounter of Cassac, 77
Close Encounters of the Third Kind, 34

competent, 6, 91
Contact, 56
control tower, 46, 47
Copely Woods Encounter, 83
Cornell, 56
Court, 3, 4, 5, 93
creatures, 9, 10, 20, 21, 57, 70, 76
Criminal courts, 5
cults, 14

Evidence, 3, 4, 5
evidentiary requirements, 5
exhibits, 7, 91
expert opinion, 7
expert witnesses, 7, 54, 92
expertise, 54, 92
extraterrestrial, 31, 54, 65, 66, 80
extremophiles, 27, 31
Ezekiel, 20, 21

D

Dallas Morning News, 41
dark matter, 58
Dechmont Woods Encounter, 82
deep sea vents, 27
demons, 37
dimensions, 39
Dinosaurs, 38
Discovery Channel, 86
disk-shaped object, 50
DNA, 3, 22, 30, 31, 94
Dominican friar, 23
Donald Shrum, 74
Dr. Alexandro Botta, 67
Drake, 3, 22, 24, 25, 26
dreams, 49, 104
dummies, 45
Dyatlov Pass Incident, 73

F

Falcon Lake Incident, 77
Family Courts, 5
Felix Moncla, 69
Fire in the Sky, 80
flares, 51, 52
flying cars, 16
flying chariots, 16
flying disk, 44, 72
flying objects, 11
Flying Saucer, 1, 48, 93, 95
flying vehicle, 16
Folsom, 42
foo fighters, 43, 63
fossil record, 38
fossils, 38
Francis Crick, 30
Frank Scully, 67

E

earth, 10, 11, 16, 19, 20, 21, 26, 27, 30, 31, 32, 38
Earth, 3, 22, 24, 26, 27, 29, 32, 49, 55, 57, 61
earth-like planets, 26
ecological niche, 38
Edward Nugent, 46
Egypt, 12, 13, 93
Egyptian, 12
Egyptians, 9, 12
electrons, 33
elements, 32, 33
Emilcin, 81
Erich von Daniken, 10
Europa, 56
European, 10, 32, 43
evidence, 5, 6, 7, 10, 34, 42, 64, 72, 83, 84, 91, 92

G

galaxies, 29, 58
galaxy, 24, 26, 29
Gazette, 18
geologic record, 31
geometric figures, 15
German, 43
German secret weapons, 43
geysers, 27
Ghost Rockets, 63
Giordano Bruno, 2, 23, 93
Giza, 12, 13, 93
Glenn Dennis, 45
gods, 14, 16, 33
gold-anodized plaque, 56
Goldilocks Zones, 28
Gordon Cooper, 54, 72, 92
Gorman Dogfight, 67
Governor Symington, 52

gravity, 50
Greeks, 32, 33
Green Fireballs, 65
Grinning Man, 76

H

habitable temperature, 28
habitable zone, 29, 107
habitable zones, 28, 29
Harbour Mille Incident, 89
Harvard, 56
helium balloon, 21
heresy, 23
hieroglyphic" images, 41
Highland, Illinois UFO, 86
historical record, 9, 17, 32
historical times, 10
Hittites, 9
homo sapiens, 38
Hopeh Incident, 63
hospitable range, 27
Hudson Valley Sightings, 83
Humanoids, 67, 74
humans, 19, 38
hunter gatherers, 38
Hynek, 34, 35, 75

I

Incas, 12
Incident At Exeter, 75
Indian, 16
Indians of the Americas, 32
Indrid Cold, 76
intelligent beings, 23, 47
Interrupted Journey, 49
Isaac Asimov, 56

J

Japan Air Lines, 3, 40, 50, 83
Japanese soldiers, 14
jellyfish, 38
jets, 47, 84, 88
Jimmy Carter's sighting, 78
John Samford, 47
Judge, 5, 7, 41, 92
Judy Doraty Abduction, 78
Julius Obsequens, 3, 8, 11

juror, 7
Jury, 5, 92

K

Kaikoura Lights, 81
Kelly Cahill, 84
Kelly Johnson/Santa Barbara Channel Case, 70
Kelly-Hopkinsville Encounter, 70
Kepler Space Mission, 26
Kepler-22, 29
Kitty Hawk, 35
knowledge, 9, 11, 14, 24, 32, 39

L

Lakenheath-Bentwaters Incident, 71
language, 37, 43, 57
languages, 12, 56, 57
legal standards, 5, 7
lenticular cloud, 70
Levelland, 72
life, 23, 24, 27, 28, 29, 30, 31, 38, 55, 56, 58, 91, 95
Livy, 11
Lonnie Zamora, 35, 74
Lubbock, 68, 72
Lubbock Lights, 68

M

magic, 37
Magic Car, 16
magical vehicles, 19
Mahabharata, 16
Mahaharata, 16
Major Jesse Marcel, 45
mammals, 38
mankind, 27, 30, 38
Mars, 31, 94, 106, 107
Mars meteorite ALH84001, 31
material, 6, 7, 14, 32, 35, 60, 91, 95
Maury Island Incident, 64
Mayans, 12
media, 34, 44, 45, 65, 87, 89, 95
medical examination, 49
Medieval, 21
Melanesian islands, 14
Meng Zhaoguo Incident, 85

ALIENS AND UFOs: CASE CLOSED / 100

meteor, 11
meteorite, 31, 48
meteorites, 38
meteoroids, 30
Meteors, 11
methane, 29
Mexico, 12, 13, 35, 44, 64, 67, 93
Michigan Swamp Gas Sightings, 75
microbial organisms, 27
Middle East, 12
Milky Way, 24, 26
mines, 27
Miracle of the Sun, 62
missile, 11
molecule, 30
molecules, 33
Monkey line drawing, 15
monuments, 12
moon, 30, 35
mother ship, 18
Mothman Prophecies, 75
Mystery Airships, 60, 62
mythology, 37

N

NASA, 26, 29, 30, 56, 84, 94, 95
nature, 27, 37, 38
Nazca, 3, 8, 10, 15, 93
Nazca Desert, 15
Nazca lines, 10
Nevada, 51, 86, 87
New World, 17
Newcastle AFB, 47
Nick Pope, 35
NJ Turnpike/Carteret Lights Incident, 87
Nuremberg, 18

O

Oakland, 42
observation, 11, 26, 32
observations, 9
oceanography, 32
Operation Mainbrace, 68
Orion, 12
Orion correlation theory, 12
Osiris, 12
oxygen, 29

P

Pacific, 14, 43, 70
panspermia, 30
Panspermia Theory, 31
Pascagoula Abduction, 79
Pentagon, 55
periodic table, 33
Peter Jennings, 86
phenomena, 9, 43
Phoenix, 3, 40, 51, 52, 86
physics, 58
pilots, 43, 54, 55, 66, 68, 87, 88
Pioneer, 56
planetoids, 30
planets, 24, 26, 29, 32, 39
plurality of inhabited worlds, 23
Plutarch, 60
Pope Pius I, 60
Portage County UFO Chase, 76
prehistory, 10
preponderance of the evidence, 5
Prescott, 51, 69
primeval soup, 30
Project Blue Book, 35, 55
Project Grudge, 34
Project Sign, 34, 66
proof, 5, 6, 7, 9, 91
protons, 33
pyramid, 12
pyramids, 10, 12

R

radar, 44, 46, 47, 55, 68, 71, 72, 79, 83, 84
radial velocity and astrometry method, 26
RAF, 63, 68, 70, 71
Rama, 16
Ramayana, 16, 20
reasonably believe, 5
relevant, 6, 91
religions, 10, 14, 58
religious rituals, 14
religious texts, 19
Renaissance, 21
Rendlesham-Woodbridge Incident, 82
researchers, 10, 14, 34
rockets, 16, 20
Roger M. Ramey,, 44

Roger Ramey, 47
Romans, 32
Rome, 11, 60
Roswell, 3, 40, 44, 45, 64, 94
Royal Canadian Air Force, 69

S

Sacramento, 42
salt flats, 27
San Francisco, 42
Santa Maria, 17
satellites, 31
saucer, 54, 63, 64, 65, 67, 73, 87
science, 32, 37, 56
Scientific American, 60
scientific thought, 23
scientists, 10, 23, 26, 27, 29, 31, 33, 38, 58
Seaweed, 38
second world war, 14
SETI, 24
sighting, 9, 17, 45, 51, 52, 60, 62, 64, 66, 69, 75, 76, 78, 82, 84, 86, 92
sightings, 9, 11, 34, 35, 37, 43, 46, 47, 51, 55, 59, 61, 63, 64, 68, 71, 73, 75, 76, 81, 85, 87, 88, 89, 91, 92
skeptic, 57
skeptics, 9, 11
sky, 10, 11, 14, 15, 18, 21, 37, 38, 42, 43, 47, 48, 51, 60, 62, 85, 87, 88
Snippy the Horse Mutilation, 77
Socorro, 35
solar system, 23, 29
Sonora, 51
sons of God, 20
South America, 12
South Pacific, 14
southern American Hemisphere, 12
Southern Illinois UFO, 86
space shuttles, 6
space vehicle, 20
space vehicles, 10, 20
Spain, 17, 80, 82
Spanish Inquisition, 17
species, 38
speed of light, 39
Spoletum, 11
St. Clair Triangle, 86
standard, 5, 6, 91
Stanford Abduction, 80

Stanton T. Friedman, 45
stars, 7, 12, 23, 24, 26, 28, 32, 35, 39, 51
Stephenville, Texas, 88
STS-48 Incident, 84
sub atomic particles, 33
Sumerian, 3, 53, 56, 57, 95
Sumerian mythology, 57
Sumerians, 9, 57
sun, 11, 16, 18, 23, 29, 32, 35, 62, 64
sunlight, 27, 28
sunlight., 27

T

Taro Cult, 14
technology, 20, 35
telescope, 23
temperature inversion, 47
Teotihuacan, 12, 13, 93
testimony, 54, 91, 92
The Flatwoods Monster, 69
Theory of Directed Panspermia, 31
thermal vents, 27
Thomas Mantell, 65
time, 9, 12, 17, 20, 23, 30, 31, 38, 39, 42, 43, 58, 72, 76, 78, 79, 88, 91, 106, 107
Tinley Park Lights, 87
Titan, 56
tolerable temperatures, 29
tracking balloon, 44
Trans-en-Provence Case, 83
transit method, 26
Travis Walton, 80
Trilobites, 38
Truman, 47, 55
Tuka Movement, 14
Tunguska Event, 61
twentieth century, 14

U

UFO, 7, 9, 10, 17, 21, 34, 35, 37, 41, 42, 43, 44, 45, 46, 49, 50, 55, 56, 59, 60, 61, 62, 63, 64, 66, 67, 68, 69, 71, 72, 73, 74, 75, 76, 77, 78, 80, 81, 82, 83, 84, 85, 86, 87, 88, 89, 90, 91, 92
UFO and Alien reports, 35, 42
UFO Hunters, 42
UFO investigator, 45

ALIENS AND UFOs: CASE CLOSED

UFOs, 1, 3, 6, 7, 8, 9, 21, 31, 33, 34, 35, 37, 40, 41, 43, 45, 46, 50, 54, 55, 57, 59, 64, 65, 66, 67, 68, 70, 75, 77, 78, 81, 82, 83, 86, 88, 89, 91, 92
unidentified flying object, 69, 76
United States, 41, 42, 44, 45, 46, 59, 61, 64, 65, 67, 68, 69, 74, 76, 95, 96
universe, 7, 23, 27, 29, 30, 31, 37, 49, 58, 91

V

Val Johnson Incident, 81
Valentich Disappearance, 81
Varese Close Encounter, 67
Venus, 56
Vimana, 16
Vimanas, 16
Voyager, 56
V-shaped, 51, 52, 86

W

warp space, 39
Washington D.C., 46, 68
Washington Flap, 47
Washington National Airport, 46, 47
water, 27, 28, 32, 42, 56

weather balloon, 44
West Freugh Incident, 71
wheel, 12
White House, 46
Whitley Strieber, 83
Wikipedea, 11
Wikipedia, 50, 54, 59, 93, 94, 95, 96
windmill, 41
witnesses, 45, 47, 51, 54, 68, 72, 74, 75, 77, 82, 84, 86, 87, 88, 91, 92
Witnesses, 3, 7, 53, 54, 68, 75
woodcut, 18
World War II, 43, 62, 66
worlds, 10, 23, 26, 27, 31, 58, 91, 107
Wright Brothers, 41
Wright-Patterson Air Force Base, 55
written accounts, 10, 21
written history, 21
written language, 15
written record, 37
written testimonies, 21
Wurtsmith AFB, 79

Z

Zecharia Sitchin, 57

The following pages are just for fun!

For readers 9 and above:

Read a *FUN* book –

Pinocchio and the Dragons of Martoon

Great to read - just for fun, and perfect for book reports. It tells a great story and speaks to today's children.

A version is available for the Kindle.

Pinocchio and the Dragons of Martoon
is a *new* Pinocchio story.

During his struggle to become a "real boy," Pinocchio travels to Martoon - a planet inhabited by warring aliens. Along with his new friends, Merlo the Martoonian, Tizzy the cat girl and Minimo the mouse - and the unwelcomed Ringmaster, he has many exciting adventures during which he learns a great deal not only about the aliens, but also about himself and his desire to become a real boy. Pinocchio and the Dragons of Martoon is an interesting blend of realism and fantasy, a charming and exciting read - with over 60 illustrations - and a book that people of all ages can enjoy.

In the classic story "The Adventures of Pinocchio" by Carlo Collodi, the carpenter Geppetto carves a block of pinewood into a boy puppet he names Pinocchio. As soon as Pinocchio's nose is carved, it begins to grow whenever he tells a lie. When his feet are carved, they kick Geppetto. At first Pinocchio is certainly not a good puppet or a good son. He needs to learn many lessons about life. But as naughty as Pinocchio is, Geppetto still loves him and gives him room to grow. Through his many adventures, failures and successes, Pinocchio learns that many things in life are not free and have to be earned.

Although the Pinocchio in this new story is carved out of wood, he looks very much like a real boy and not at all like the

Pinocchio drawn or imagined by artists of other books, films and television. One would have to look closely at this Pinocchio to notice the fine grains of polished wood. He also looks and speaks like a real boy, just like the one he dreams of becoming. In his quest for this dream, Pinocchio has many adventures, including "Pinocchio and the Dragons of Martoon", which starts when both he and Geppetto were swallowed by the whale.

About the author

Angelo Tropea is the author of more than two dozen non-fiction and fiction books. "Pinocchio and the Dragons of Martoon" is his first Pinocchio story.

He has just completed the sequel, "Pinocchio's Scary Adventure," and is currently writing the third book in the Pinocchio series, "Pinocchio and the Time Portal."

Names of persons

The following names of persons are offered without lengthy editorial comment (and in random order) in the hope that you will use them as a key to open many strange and wondrous doors...

Isaac Asimov
Frank Drake
Carl Sagan
Eric von Daniken
George Lucas
Steven Spielberg
Anunnaki
Philip Coppens
Professor Stephen Hawking
Jimmy Carter
Giordano Bruno
Dr. Josef Allen Hynek
Edgar Mitchell
Bob Lazar
Gordon Cooper
Giorgio A. Tsoukalos
John Lennon
Alexander the Great
Philip J. Corso
Edmund Halley
Betty and Barney Hill
Jules Verne
Professor Francis Crick
Stanton Friedman
Kenneth Arnold
Plutarch
Lonnie Zamora
Jules Verne
Ezekiel
Zecharia Sitchin

Names of places

Places are not only physical locations, but also states of time and mind...

Area 51
Triangle Over Phoenix
Roswell
Europa
Holloman Air Force Base
Area 51 (again)
Callisto
Aurora
Garden of Eden
Mars
Titan
Kecksburg, Pennsylvania
Enceladus
Orion Nebula
Easter Island
Bermuda Triangle
Stonehenge
Orion
Bermuda Triangle
Groom Lake
Exeter
Sirius
Sumeria
Belgium
Olympus
Wright-Patterson AFB
Las Vegas
Rendlesham Forest
White Sands Proving Grounds
Brooklyn, New York

ALIENS AND UFOs: CASE CLOSED / 107

Names of things

Some things are universal and belong to all...

flying saucer
UFO
exoplanets
Meteorites
V-shape
DNA
Voyager Spacecraft
SETI
Kepler Telescope
missing time
Panspermia theory
interdimensional
evolution
close encounter
pyramids
habitable zone
abduction
plurality of worlds
ALH84001
Directed Panspermia theory
MUFON
alien abduction
"When Worlds Collide"
Helium 3
Cattle mutilation
Reverse engineering
NASA Mars Rover
the speed of light
grey alien
Majestic
Big bang

"On November 24, 1992, a UFO crashes in Southaven Park, Shirley, NY. John Ford, a Long Island MUFON researcher, investigates the crash. On June 12, 1996, Ford is arrested and charged with plotting to poison several local politicians by sneaking radium in their toothpaste.

On advice of counsel Ford pleads insanity and is committed to the Mid Hudson Psychiatric Center. Critics say the charges are a frame-up."

from the Wikipedia article "UFO conspiracy theory"

ALIENS AND UFOs: CASE CLOSED / 109

LET YOUR CURIOSITY SOAR....

FASTER THAN THE SPEED OF LIGHT!

I TOLD MY WIFE ONCE THAT I WAS AN ALIEN (I WAS JUST KIDDING THEN)...

ALIENS AND UFOs: CASE CLOSED / 112

NOW I'M NOT SO SURE.

ALIENS AND UFOs: CASE CLOSED / 113

The following two pages contain a
"List of reasons why Aliens and UFOs cannot possibly exist"

ALIENS AND UFOs: CASE CLOSED / 114

"List of reasons why Aliens and UFOs cannot possibly exist"
(continued)

ALIENS AND UFOs: CASE CLOSED / 115

Printed in Great Britain
by Amazon

69007865R00068